THE BIODYNAMIC SOWING AND PLANTING CALENDAR

2009

The original biodynamic sowing and planting calendar showing the optimum days for sowing, pruning and harvesting various plant-crops, as well as for beekeeping

Compiled by Maria and Matthias Thun

Floris Books

Translated by Bernard Jarman
Additional astronomical material
by Wolfgang Held and Christian Maclean

Published in German under the title *Aussaattage*.
English edition published by Floris Books

British Library CIP Data available

ISBN 978-086315-611-3
ISSN: 1751-0449

Produced in Poland

Walther Thun, Autumn in Ederbergland, *(watercolour, 50 x 67 cm)*

Foreword

In this year's calendar we have decided not to focus as much on plant cultivation as we usually do. Instead we are trying to address some of the many questions which readers have sent to us.

The large number of natural disasters that occurred during the last year unleashed a torrent of questions from our readers about life and world events. We cannot answer all these questions individually. Instead we have tried to address those we believe to be especially important and include them as paragraphs within these pages. We hope that by doing so we will contribute towards developing a wider and deeper understanding for planetary influences.

The contrasting springs of 2007 and 2008

Following cold wintry weather during January, February and March of 2007, April became unseasonably warm. Venus already entered Aries during March but was unable to alter the wintry weather. The second half of April became exceptionally warm with daytime temperatures often reaching 30°C (86°F). However, there was no rain and the seeds which were already sown could not germinate. The first rain only came on May 7.

During the spring of 2008 by contrast we experienced night frosts right through to the end of April. The days were overcast and frequent rain meant that the soil was too wet to cultivate. On April 17 Mercury entered Aries. Usually this constellation brings warmth to the soil and warmer nights. This year, however, it was different due apparently to Jupiter being at its descending node. This caused the weather through-out April to be unpredictable.

Many of our readers wrote and asked when it would finally become warm. I comforted them with the promise of change on May 1 — a Flower day with two Warmth trines. It did indeed become warmer with daytime temperatures reaching between 24° and 28°C (75°–82°F) although it was still very windy. A few more days were needed before we could cultivate the fields with our implements. We have never had such a late start to our spring work.

The first ten days of May brought humanitarian disasters too, with the tidal waves and earthquakes in Burma and China. These were stimulated by difficult Pluto-Neptune aspects.

The combined effect of unfavourable Mercury and Pluto aspects with those of the Sun and Pluto set the scene for the catastrophic storms which struck Burma causing tens of thousands of deaths. An occultation (the Moon being directly infront) of Neptune combined with an unfavourable aspect of the Sun to Neptune set off the earthquake in China, again with tens of thousands of deaths.

An aspect encouraging volcanic activity stimulated new eruptions of Mount Etna in Sicily. These natural disasters are usually caused by or at least occur contemporaneously with the negative positions of the classical planets in relation to Neptune and Pluto.

In memory of Martin Bender

Martin Bender grew up on a farm in Hungary. After the Second World War all German settlers were expelled and so he came to Dexbach. He worked here in an iron foundry until he retired. He had had a great love for farming and the care of plants and animals ever since he was a child and now came to offer his time to the farm. He helped us on the farm and with our planting trials for a further twelve years.

He stirred the preparations for us hundreds of times and did so with great devotion. He often questioned if the barrel preparation should really only be stirred for twenty minutes or whether it might be more effective if stirred for an hour like horn manure or horn silica.

For many weeks during the summer he spent his time weeding and often wondered why we seemed to have more weeds than he remembered having in Hungary. He was so quick with his hands that no other co-worker could keep pace with him. Early each morning he announced what he planned to do and was usually finished much sooner than was expected. He was always polite and contented.

In November 2007 after he had swept up the last leaves in the courtyard he became unwell. He comforted me, "I will be well again by the time next year's leaves need gathering." His health, however, got worse and in March 2008 he passed over the threshold of death. We miss him keenly as we work and remember him warmly and full of gratitude.

Martin Bender stirring preparations ...   *and making compost*

Note for new readers of the Calendar

For over 56 years we have been researching the influences of cosmic rhythms and constellations on agriculture, gardening and in a specific way on insect life. The results of this research provide the basis for the recommendations given each year in the Calendar. Because we cannot fit all the new results into a small booklet like this, we have produced a number of books that bring together important information gathered over a long period of time and include detailed descriptions of the trials referred to. In English *The Results of the Biodynamic Sowing and Planting Calendar* contain a lot of these details.

When I began sowing radishes in 1952 I found great variation in the growth patterns of the plants sown over a period of ten days. This inspired me to sow radishes under the same growing conditions during a period of several weeks.

Because I could not explain the differences and was as yet unaware of the influences affecting plant growth, I continued my work into the following year. I thought that the differences might be explained by the rhythmical movements in the cosmos and so decided to study astronomy in more depth.

After a few years I found that differences were most marked when cultivation and sowing took place on the same day. If I prepared a large seed bed about ten metres (33 feet) long and sowed it daily with a row of radishes, I found that the differences between them were much smaller than if the ground had been prepared fresh each day.

There were big differences in the developing leaf forms. If, however, I watered the plants during dry weather, I found that all the new leaves had the same form. These were two key experiences to be considered if I was to find out what lay behind these results.

After nine years the first results of my work could be published. In the meantime I had learnt that if I cultivated the soil to a spade's depth, cosmic influences would be activated in the soil. They could then be accessed by the newly sown seeds and express themselves in the forms of the developing plant. These influences originate in the constellations of the zodiac and are mediated to the earth by the Moon. The Moon's effect is helped by the classical elements earth, water, air/light and warmth. Since these elements have their origin in the constellations, the passage of the moon through the associated signs of the zodiac enables different growth impulses to find expression when seeds are sown.

As time went by we discovered further cosmic effects. These came about through the movement of the planets and also affected plant growth.

Along with impulses which are favourable to plant growth are those which have a detrimental effect. Such unfavourable sowing times include those which encourage premature pest attack or produce infertile seeds.

We have also found in relation to the biodynamic preparations, that if they are applied during favourable constellations, they will stimulate growth. If however they are applied during unfavourable periods, they can damage the plant, hold back growth and reduce quality.

Included in the planting calendar are recommendations for the planting, care and cultivation of crops as well as the application of biodynamic preparations. By working in harmony with cosmic intentions good yields of the best quality can be achieved.

A willow twig

On March 28, 1966 the Sun aspected 135° to Neptune and on March 29 Venus was 90° and Mars 144° to Neptune. During the night there was a blizzard and temperatures fell to −4°C (25°F). When I went outside in the morning I found, beneath the willow tree, a small willow twig about 45 cm long. I took it indoors and put it in some water. After some weeks I found that it had grown roots. On May 23 I drove to our bee house in Dexbach. I took the little willow plant with me and planted it out.

This is where we intended to carry out our planting trials, but to do this we needed water. A certain Mr Ortwein helped by dowsing for water. He found a source deep in the ground some three metres from the little willow plant. And so at the beginning of November with the help of Aunt Gertrud, he began digging a well two metres in diameter. Two and a half metres down they found the water.

A large pipe was installed along with a pump to bring the water up and a block of wood was put on top to cover it. This has since been regularly used as our breakfast table and we made some wooden seats too and put them in a circle. When working with our farm team we could then take a break and sit in the shade beneath the willow which has shared in so many of our joys and sorrows. Forty-two years later its trunk has attained a diameter of 260 cm and we are grateful each year for the shade provided by its far spreading branches.

The forty-two-year-old willow tree

What are oppositions, trines and conjunctions?

Opposition ☍
A geocentric opposition occurs when for the observer on the Earth there are two planets opposite one another — 180° apart — in the heavens. They look at one another from opposite sides of the sky and their light interpenetrates. Their rays fall on to the earth and stimulate in a beneficial way the seeds that are being sown in that moment. In our trials we have found that seeds sown at times of opposition resulted in a higher yield of top quality crops.

With a *heliocentric* opposition an observer would need to place himself on the sun. This is of course physically impossible but we can think our way to an understanding of it. The sun is in the centre and the two planets placed 180° apart also gaze at each other but this time across the circle of the Sun's orbit. Their rays are also felt by the Earth and stimulate better plant growth.

At times of opposition two zodiac constellations are also playing their part. If one planet is standing in a Warmth constellation, the second one will be in a Light constellation or vice versa. If one planet is in a Water constellation, the other will be in an Earth one.

The Trine △
Trines occur when planets are 120° from one another. The two planets are then both standing in the same elemental configuration — Aries and Leo for example are both Warmth constellations. A Warmth trine means that the effects of these constellations will enhance fruit and seed formation in the plants sown at this time. If two planets are in trine position in Water, watery influences will be enhanced which usually brings high rainfall. Plants sown on these days will yield more leaf than those on other days. Trine effects can change the way plants grow.

Conjunctions ☌
Conjunctions and multiple conjunctions occur when two or more planets stand behind one another in space. It is then usually only the planet closest to the earth, which has any influence on plant growth. If this influence is stronger than that of the sidereal moon cosmic disturbances can occur that irritate the plant and cause checks in growth. This negative effect is increased further when the Moon or another planet stands directly in front of another — an occultation (☉) or eclipse in the case of Sun and Moon). Sowing at these times will affect subsequent growth detrimentally and harm a plant's regenerative power.

The zodiac

The **zodiac** is a group of twelve constellations of stars which the Sun, Moon and all the planets pass on their circuits. The Sun's annual path always takes exactly the same line called **ecliptic.** The Moon's and planet's paths vary slightly, sometimes above and sometimes below the ecliptic. The point at which their paths cross the ecliptic is called a **node** (☊ and ☋).

If New Moon is at a node there is a **solar eclipse,** as the Moon is directly in front of the Sun, while Full Moon at a node causes a **lunar eclipse** where the Earth's shadow falls on the Moon. If Sun or Moon pass exactly in front of a planet there is an **occultation** (•). If Mercury or Venus pass exactly in front of the Sun, this is a **transit** (other planets cannot pass in front of the Sun).

The angles between Sun, Moon and planets are called aspects. In this calendar the most important is the 120° angle, or **trine.**

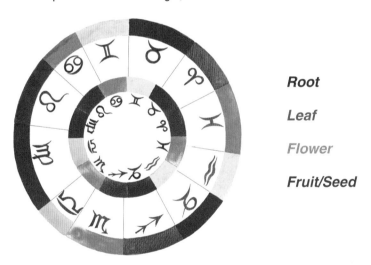

Root

Leaf

Flower

Fruit/Seed

In this illustration the outer circle shows the varying sizes of the visible **constellations** of the **zodiac.** The dates on this outer circle are the days on which the Sun enters the constellation (this can change by one day because of leap years). The inner circle shows the divisions into equal sections of 30° corresponding to the **signs** used in astrology.

It is the *constellations* on which our observations are based, and which are used throughout this calendar.

Trines △

The twelve constellations are grouped into four different types, each having three constellations at an angle of about 120°, or trine. About every nine days the Moon passes a similar region of forces. Through hoeing or spraying silica (501) in the rhythm of the trines we can reawaken the impulse of the original sowing day.

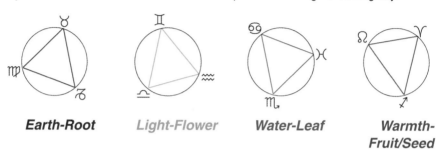

Earth-Root   *Light-Flower*   **Water-Leaf**   **Warmth-Fruit/Seed**

The sidereal Moon

In its 27-day orbit round the Earth the Moon passes the twelve regions of the zodiac and transmits forces to the Earth which find expression in the four elements. They bring about enlivening processes in the four different plant organs. At the point of sowing, cultivation and harvest we can stimulate the growth and health of a plant.

These cosmic forces work similarly in bees and can be used to enhance their life. The colony shuts itself away from the outer world in the skep or box by sealing it with propolis. When we open the skep it causes a certain 'confusion' in the colony. A new cosmic impulse can work into this unrest, directing the activity of the bees until the next routine disruption.

Let us summarize the laws pertaining to plant-experiment, beekeeping and weather observation in one table (opposite).

The different impulses vary between two and four days. This basic framework is sometimes interrupted. It can happen that planetary oppositions override some days with their different impulses. It may be also that trine positions activate a different element to the one the Moon is transmitting on this day. Days when either the Moon's path intersects with the ecliptic (ascending ☊ or descending ☋ node) produce mainly negative effects which are intensified if there is an eclipse or occultation during which the influence of the more distant planet is interrupted or changed by the nearer one. Such times are unsuitable for sowing and harvesting.

Constellation	Sign	Element	Weather	Plant	Bees
Pisces, Fishes	♓	W Water	Watery	Leaf	Making honey
Aries, Ram	♈	H Warmth	Warm/hot	Fruit	Gathering nectar
Taurus, Bull	♉	E Earth	Cool/cold	Root	Building comb
Gemini, Twins	♊	L Light	Airy/bright	Flower	Gathering pollen
Cancer, Crab	♋	W Water	Watery	Leaf	Making honey
Leo, Lion	♌	H Warmth	Warm/hot	Fruit	Gathering nectar
Virgo, Virgin	♍	E Earth	Cool/cold	Root	Building comb
Libra, Scales	♎	L Light	Airy/bright	Flower	Gathering pollen
Scorpio, Scorpion	♏	W Water	Watery	Leaf	Making honey
Sagittarius, Archer	♐	H Warmth	Warm/hot	Fruit	Gathering nectar
Capricorn, Goat	♑	E Earth	Cool/cold	Root	Building comb
Aquarius, Waterman	♒	L Light	Airy/bright	Flower	Gathering pollen

Groupings of plants for sowing and harvesting

In cultivating a plant, particular parts are developed for food. We can divide them into four groups.

Root crops on Root Days
Development in the root region is in radishes, swedes, sugar beet, beetroot celeriac, carrot, scorzonera, etc. Potatoes and onions are included in this group too. These days produce good yields and top storage quality.

Leaf plants on Leaf Days
Development in the leaf realm is in the cabbage family, lettuce spinach, lambs lettuce, endive, parsley, leafy herbs and fodder plants. Leaf days are suitable for sowing and tending these plants but not for harvesting and storage. For these (as well as harvesting of cabbage for sauerkraut) Fruit and Flower days are recommended.

Flower plants on Flower Days
These days are favourable for sowing and tending all kinds of flower plants but also for cultivating and spraying 501 on oil-bearing plants such as linseed, rape, sunflower, etc. Cut flowers have the strongest scent and remain fresh for longer if cut on Flower days, and the mother plant brings forth many new side shoots. If flowers for drying are harvested on Flower days they retain the most vivid colours. If cut on other days they soon lose their colour. Oil-bearing plants are best harvested on Flower days.

Fruit Plants on Fruit Days

To this category belong all those plants which have their 'fruit' in the realm of the seed such as beans, peas, lentils, soya, maize, tomatoes, cucumber, pumpkin, courgettes, but also cereals for summer and winter crops. Sowing oil-bearing plants brings the best yields of seeds. The best time for extraction of oil later on is on Flower days. To grow good seed, Leo days are particularly suitable. Fruit plants are best harvested on Fruit days. They promote a good quality of storage and support regeneration. If storing fruit additionally choose the time of the ascending Moon.

There is always some uncertainty as to which category some plants belong. Onions and beetroot give a similar yield when sown on Root and Leaf days, but the keeping quality is best from Root days. Kohlrabi and cauliflowers belong to Leaf days, as does Florence fennel. Broccoli is more beautiful and firmer when sown on Flower days.

Explanations of the calendar pages

Next to the date is the constellation (and time of entry) in which the Moon is. This is the astronomical constellation, not the astrological sign, see page 9. In the next column solar and lunar events are indicated.

A further column shows which element works predominantly on that day (this is useful for beekeepers). Note **H** is used for warmth (heat). Sometimes there is change during the day when both elements are mentioned in order. Warmth effects on thundery days are implied but are not mentioned in this column, but may have a ⟨ symbol in the far right.

The next column shows in colour the part of the plant which will be enhanced by sowing or cultivation on that day. Numbers indicate times of day. On the extreme right special events in nature are noted as well as anticipated weather changes which disturb or break up the overall weather pattern.

When elements are indicated which do not correspond to the Moon's position in the zodiac (sometimes it may even be more than one element on the same day), it is not a misprint, but takes account of other cosmic aspects which overrule the Moon-zodiac pattern and have an effect on a different part of the plant.

Unfavourable times are marked thus (- - - -). These are caused by eclipses, nodal points of the Moon or the planets or other aspects with a negative influence; they are not elaborated in the Calendar. If one has to sow at unfavourable times for practical reasons, one can choose favourable days for hoeing and so bring about an improvement.

On the opposite page astronomical aspects are indicated (see Astronomical Symbols below) with those visible to the naked eye shown in **bold** type. Visible conjunctions (particularly those involving Mercury) are not always visible from all parts of the Earth. SH indicates visible in southern hemisphere only.

Astronomical symbols

Constellations
- ♓ Pisces
- ♈ Aries
- ♉ Taurus
- ♊ Gemini
- ♋ Cancer
- ♌ Leo
- ♍ Virgo
- ♎ Libra
- ♏ Scorpio
- ♐ Sagittarius
- ♑ Capricorn
- ♒ Aquarius

Planets
- ☉ Sun
- ☾,☽ Moon
- ☿ Mercury
- ♀ Venus
- ♂ Mars
- ♃ Jupiter
- ♄ Saturn
- ♅ Uranus
- ♆ Neptune
- ♇ Pluto
- ● Full Moon
- ● New Moon

Other symbols

Aspects
- ☊ Ascending node
- ☋ Descending node
- ⌒ Highest Moon
- ⌣ Lowest Moon
- **Pg** Perigee
- **Ag** Apogee
- ☍ Opposition
- ☌ Conjunction
- ⸰ Eclipse/occultation
- ⸰ Lunar eclipse
- △ Trine

- **St** Storms likely
- **ᛡ** Thunder likely
- **Eq** Earthquakes
- **Tr** Traffic dangers
- **Vo** Volcanic activity

Northern Planting Time

Southern Planting Time

E Earth L Light/Air W Water H Warmth/Heat

Two-weekly planting times

The ascending and descending Moon

From midwinter through to midsummer the Sun rises earlier and sets later each day while its path across the sky ascends higher and higher. From midsummer until midwinter this is reversed, the days get shorter and the midday Sun shines from an ever lower point in the sky. This annual ascending and descending of the Sun creates our seasons in both the northern and southern hemispheres. The difference is that while it is ascending (spring) in the north it is descending (autumn) in the south and vice versa. As it ascends and descends during the course of the year the Sun is slowly moving, from an earth-centred point of view, through each of the twelve constellations of the zodiac in turn. It shines for approximately one month from each constellation.

In the northern hemisphere the winter solstice occurs when the Sun is in Sagittarius and the summer solstice when it is Gemini. At any point between Sagittarius and Gemini the Sun is ascending. Likewise between Gemini and Sagittarius it is descending. In the southern hemisphere this is reversed.

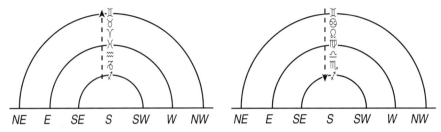

Northern hemisphere ascending Moon (left) and descending Moon: Planting Time

The Moon and all the planets follow approximately the same path as the Sun around the zodiac but while the Sun takes twelve months to complete the circuit the Moon takes about 27$^1/_2$ days to complete one rotation. This means that the Moon will ascend for about fourteen days and then descend.

Each month the Moon completes one cycle of the zodiac and shines from each constellation in turn for a period of two to three days. It is important to distinguish the journey of the Moon through the zodiac (siderial rhythm) from the waxing and waning (synodic) cycle. In any given constellation there may be a waxing, waning, full, quarter, sickle or gibbous Moon. As it moves through the zodiac the Moon, like the Sun, is ascending (in the northern hemisphere) when it is in the constellations from Sagittarius to Gemini and descending between Gemini and Sagittarius. In the southern hemisphere it is ascending between Gemini and Sagittarius and descending between Sagittarius and Gemini.

During the ascending Moon period plant sap rises more strongly. The upper part of the plant fills with sap and vitality. This is a good time for cutting scions (for grafting). Fruit harvested during this period remains fresh for longer when stored.

During the descending Moon period plants take root readily and connect well with their new location. Moving plants from one location to another is termed transplanting. This is the case when young plants are moved out from the seed bed into their final growing position but also when the gardener wishes to strengthen the root development of young fruit trees, shrubs or pot plants by frequently re-potting them. Sap movement is slower during the descending Moon. This is why it is good time for trimming hedges, pruning trees and felling timber as well as applying compost to meadows, pastures and orchards.

Note that *sowing* is the moment when a seed is put into the soil; either the ascending or descending period can be used. It then needs time to germinate and grow. This is different from *transplanting* which is best done during the descending Moon. These times given in the calendar. Northern Planting Times refer to the northern hemisphere, and **Southern Planting Times** refer to the southern hemisphere. All other constellations and planetary aspects are equally valid in both hemispheres.

Local times

Times given are *Greenwich Mean Time,* using 24-hour clock with ^{h} after the time. Thus 15^h is 3 pm. **No account is taken of daylight saving (summer) time (*DST*).** Note 0^h is midnight at the beginning of a date, and 24^h is midnight at the end of the date.

For different countries adjust as follows:

Britain, Ireland, Portugal, Iceland: GMT (DST from March 29 to Oct 24, add 1^h)
Central Europe: add 1^h (DST from March 29 to Oct 24, add 2^h)
Eastern Europe (Finland, Baltic states, etc.): add 2^h (DST from March 29 to Oct 24, add 3^h)

Namibia: add 1^h (DST to April 4 and from Sep 6, add 2^h)
Israel: add 2^h (DST from March 27 to Sep 19, add 3^h)
Egypt: add 2^h (DST probably from May 1 to Sep 24, add 3^h)
South Africa: add 2^h (no DST)   *Kenya:* add 3^h (no DST)
Pakistan: add 5^h (no DST)   *India:* add $5^{1}/_{2}^{h}$ (no DST)
Bangladesh: add 6^h (no DST)

Western Australia: add 8^h (DST to March 28 and perhaps from Oct 25, add 9^h)
South Australia: add $9^{1}/_{2}^{h}$ (DST to April 4 and from Oct 4, add $10^{1}/_{2}^{h}$)
Northern Territory: add $9^{1}/_{2}^{h}$ (no DST)   *Queensland:* add 10^h (no DST)
ACT, NSW, Victoria, Tasmania: add 10^h (DST to April 4 and from Oct 4, add 11^h)
New Zealand: add 12^h (DST to April 4 and from Sep 27, add 13^h)

Argentina: subtract 3^h (no DST)
Brazil (Eastern): subtract 3^h (DST to Feb 21 and from Oct 11, subtract 2^h)
Chile: subtract 4^h (DST to March 14 and from Oct 11, subtract 3^h)
Newfoundland Standard Time: subtract $3^{1}/_{2}^{h}$ (DST March 8 to Oct 31, subtract $2^{1}/_{2}^{h}$)
Atlantic Standard Time: subtract 4^h (DST March 8 to Oct 31, subtract 3^h)
Eastern Standard Time: subtract 5^h (DST, March 8 to Oct 31, subtract 4^h)
Central Standard Time: subtract 6^h (DST, except *Sask.* March 8 to Oct 31, subtract 5^h)
Mexico (mostly CST): subtract 6^h (DST, April 5 to Oct 24, subtract 5^h)
Mountain Standard Time: subtract 7^h (DST, except *AZ,* March 8 to Oct 31, subtract 6^h)
Pacific Standard Time: subtract 8^h (DST March 8 to Oct 31, subtract 7^h)
Alaska Standard Time: subtract 9^h (DST March 8 to Oct 31, subtract 8^h)
Hawaii Standard Time: subtract 10^h (no DST)

January 2009

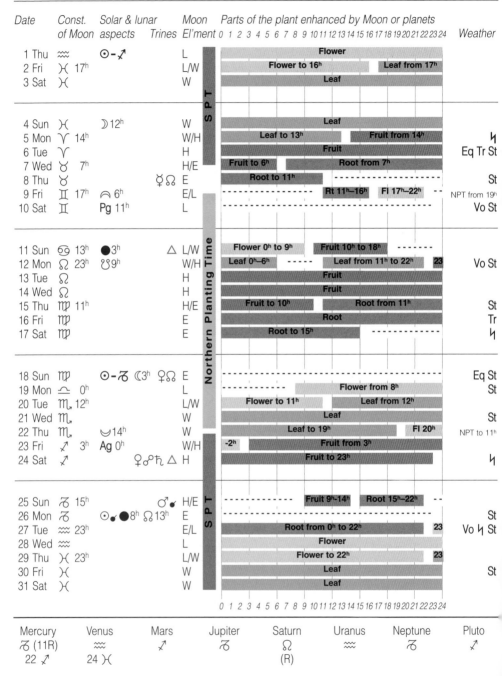

Date	Const. of Moon	Solar & lunar aspects	Trines	Moon El'ment	Parts of the plant enhanced by Moon or planets	Weather
1 Thu	♒︎	☉–♐		L	Flower	
2 Fri	♓ 17ʰ			L/W	Flower to 16ʰ / Leaf from 17ʰ	
3 Sat	♓			W	Leaf	
4 Sun	♓	☽ 12ʰ		W	Leaf	
5 Mon	♈ 14ʰ			W/H	Leaf to 13ʰ / Fruit from 14ʰ	♄
6 Tue	♈			H	Fruit	Eq Tr St
7 Wed	♉ 7ʰ			H/E	Fruit to 6ʰ / Root from 7ʰ	
8 Thu	♉		☿♌	E	Root to 11ʰ	St
9 Fri	♊ 17ʰ	♉ 6ʰ		E/L	Rt 11ʰ–16ʰ Fl 17ʰ–22ʰ	NPT from 19ʰ
10 Sat	♊	Pg 11ʰ		L		Vo St
11 Sun	♋ 13ʰ	● 3ʰ	△	L/W	Flower 0ʰ to 9ʰ / Fruit 10ʰ to 18ʰ	
12 Mon	♌ 23ʰ	☍ 9ʰ		W/H	Leaf 0ʰ–6ʰ / Leaf from 11ʰ to 22ʰ 23	Vo St
13 Tue	♌			H	Fruit	
14 Wed	♌			H	Fruit	
15 Thu	♍ 11ʰ			H/E	Fruit to 10ʰ / Root from 11ʰ	St
16 Fri	♍			E	Root	Tr
17 Sat	♍			E	Root to 15ʰ	♄
18 Sun	♍	☉–♑ ☾ 3ʰ	♀♌	E		Eq St
19 Mon	♎ 0ʰ			L	Flower from 8ʰ	St
20 Tue	♏ 12ʰ			L/W	Flower to 11ʰ / Leaf from 12ʰ	
21 Wed	♏			W	Leaf	St
22 Thu	♏	◡ 14ʰ		W	Leaf to 19ʰ / Fl 20ʰ	NPT to 11ʰ
23 Fri	♐ 3ʰ	Ag 0ʰ		W/H	-2ʰ / Fruit from 3ʰ	
24 Sat	♐		♀☌♄ △	H	Fruit to 23ʰ	♄
25 Sun	♑ 15ʰ		♂☌	H/E	Fruit 9ʰ–14ʰ Root 15ʰ–22ʰ	
26 Mon	♑	☉☌ ● 8ʰ ☍13ʰ		E		St
27 Tue	♒︎ 23ʰ			E/L	Root from 0ʰ to 22ʰ 23	Vo ♄ St
28 Wed	♒︎			L	Flower	
29 Thu	♓ 23ʰ			L/W	Flower to 22ʰ 23	
30 Fri	♓			W	Leaf	St
31 Sat	♓			W	Leaf	

Northern Planting Time

Mercury	Venus	Mars	Jupiter	Saturn	Uranus	Neptune	Pluto
♑ (11R)	♒︎	♐	♑	♌	♒︎	♑	♐
22 ♐	24 ♓			(R)			

NB: All zodiac symbols refer to astronomical constellations, not astrological signs (see p.9)

Planetary aspects

*(**Bold** = visible to naked eye)*

1
2 ☽☌♅ 14^h ☽☍♄ 18^h
3
4
5
6
7
8
9
10 ☽☍♂ 16^h
11 ☉△♄ 18^h ☾☍♃ 20^h
12 ☾☍☿ 6^h
13 ☾☍♆ 7^h
14 ☾☍♀ 14^h
15 ☾☍♅ 4^h **☾☌♄ 8^h**
16
17
18 ☿☌♃ 19^h
19
20 ☉☌☿ 16^h
21
22
23 ♀☌♅ 1^h
24 ☉☌♃ 6^h ♂△♄ 8^h ♀☍♄ 8^h
25 ☾●♂ 3^h ☾☌☿ 9^h
26 ☾☌♃ 5^h
27 ☿☌♂ 6^h ☽☌♆ 17^h
28
29 ☽☌♅ 21^h ☽☍♄ 23^h
30 **☽☌♀ 9^h**
31

Planet (naked eye) visibility
Evening: Mercury, Jupiter (to mid month), Venus
All night: Saturn
Morning:

January 2009

Mercury, Jupiter and Neptune are in Capricorn which could lead to a proper winter. However, Mars, Saturn and Pluto are in the Warmth constellations of Leo and Sagittarius and whether they allow this to occur is an open question. Venus follows her own course through the Light region of Aquarius, moving into the spring constellation of Piseces in the last week of January, This usually stimulates the appearance of snowdrops and other early spring flowers. This is particularly likely due to the opposition with Saturn. Five planetary conjunctions and three occultations influence the month of January.

Northern Planting Time
Jan 9 19^h to Jan 22 11^h.
Southern Planting Time
Dec 27 to Jan 7 22^h and Jan 22 17^h to Feb 5.

Planting time is also suitable for pruning **fruit trees, vines** and **hedges,** choosing Fruit days for fruiting and flowering plants.

When **milk processing** it is best to avoid unfavourable times (- - - - -). This applies to both butter and cheese making. Milk, which has been produced on Warmth/Fruit days, yields the highest butterfat content. This is also the case on days with a tendency for thunderstorms. Times of moon perigee (**Pg**) are almost always unfavourable for milk processing and even yoghurt will not turn out well. Starter cultures from such days decay rapidly and it is advisable to produce double the amount the day before. Milk loves Light and Warmth days best of all. Water days are unsuitable.

February 2009

All times in GMT

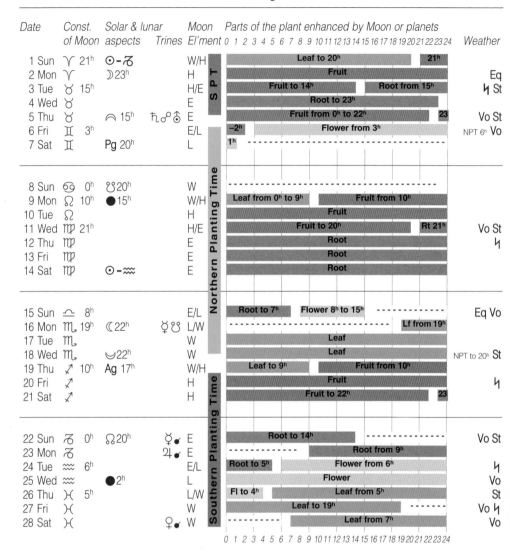

Date	Const. of Moon	Solar & lunar aspects	Trines	Moon El'ment	Planting time	Parts of the plant enhanced by Moon or planets (0–24)	Weather
1 Sun	♈ 21^h	☉-♑		W/H	SPT	Leaf to 20^h · 21^h	
2 Mon	♈	☽23^h		H	SPT	Fruit	Eq
3 Tue	♉ 15^h			H/E	SPT	Fruit to 14^h · Root from 15^h	♄ St
4 Wed	♉			E	SPT	Root to 23^h	
5 Thu	♉	⩀ 15^h	♄☍⛢	E	SPT	Fruit from 0^h to 22^h · 23	Vo St
6 Fri	♊ 3^h			E/L	Northern Planting Time	–2^h · Flower from 3^h	NPT 6^h Vo
7 Sat	♊	Pg 20^h		L	Northern Planting Time	1^h - - - - - - - -	
8 Sun	♋ 0^h	☋20^h		W	Northern Planting Time	- - - - - - - -	
9 Mon	♌ 10^h	●15^h		W/H	Northern Planting Time	Leaf from 0^h to 9^h · Fruit from 10^h	
10 Tue	♌			H	Northern Planting Time	Fruit	
11 Wed	♍ 21^h			H/E	Northern Planting Time	Fruit to 20^h · Rt 21^h	Vo St
12 Thu	♍			E	Northern Planting Time	Root	♄
13 Fri	♍			E	Northern Planting Time	Root	
14 Sat	♍	☉-♒		E	Northern Planting Time	Root	
15 Sun	♎ 8^h			E/L	Northern Planting Time	Root to 7^h · Flower 8^h to 15^h - - - - - -	Eq Vo
16 Mon	♏ 19^h	☾22^h	☿☋	L/W	Northern Planting Time	- - - - - - - - Lf from 19^h	
17 Tue	♏			W	Northern Planting Time	Leaf	
18 Wed	♏	◡22^h		W	Northern Planting Time	Leaf	NPT to 20^h St
19 Thu	♐ 10^h	Ag 17^h		W/H	Southern Planting Time	Leaf to 9^h · Fruit from 10^h	
20 Fri	♐			H	Southern Planting Time	Fruit	♄
21 Sat	♐			H	Southern Planting Time	Fruit to 22^h · 23	
22 Sun	♑ 0^h	☊20^h	☿•	E	Southern Planting Time	Root to 14^h - - - - - -	Vo St
23 Mon	♑		♃•	E	Southern Planting Time	- - - - - - Root from 9^h	
24 Tue	♒ 6^h			E/L	Southern Planting Time	Root to 5^h · Flower from 6^h	♄
25 Wed	♒	●2^h		L	Southern Planting Time	Flower	Vo
26 Thu	♓ 5^h			L/W	Southern Planting Time	Fl to 4^h · Leaf from 5^h	St
27 Fri	♓			W	Southern Planting Time	Leaf to 19^h - - - -	Vo ♄
28 Sat	♓		♀•	W	Southern Planting Time	- - - - - Leaf from 7^h	Vo

Mercury	Venus	Mars	Jupiter	Saturn	Uranus	Neptune	Pluto
♐ (1D)	♓	♐	♑	♌	♒	♑	♐
12 ♑		2 ♑		(R)	12 ♓		

NB: All zodiac symbols refer to astronomical constellations, not astrological signs (see p.9)

Planetary aspects
*(**Bold** = visible to naked eye)*

1
2
3
4
5 ♄♂♅ 12ʰ
6
7 ☽♂☿ 19ʰ

8 ☽♂♂ 9ʰ ☽♂♃ 18ʰ
9 **☽♂♆ 20ʰ**
10
11 **☽♂♄ 16ʰ** **☽♂♅ 17ʰ**
12 ☉♂♆ 13ʰ **☽♂♀ 21ʰ**
13
14

15
16
17 ♂♂♃ 16ʰ
18
19
20
21

22 **☽♂☿ 22ʰ**
23 **☽♂♃ 1ʰ** **☽♂♂ 7ʰ**
24 **☽♂♆ 2ʰ** ☿♂♃ 7ʰ
25
26 ☽♂♄ 1ʰ ☽♂♅ 6ʰ
27
28 **☽♂♀ 0ʰ**

February 2009

For the first ten days Mercury is retrograde in Sagittarius and will bring increased warmth influences during this period. On Feb 12 it moves into Capricorn joining Mars which has been in this constellation since Feb 2. Four planets are now in the Cold region of Capricorn and we must reckon with genuine winter weather in the northern hemisphere, especially since the Sun is also in this area until the middle of the month.

Northern Planting Time
Feb 6 6ʰ to Feb 18 20ʰ.
Southern Planting Time
Jan 22 to Feb 5 9ʰ and Feb 20 0ʰ to March 4.

The Saturn-Uranus opposition on Feb 5 is particularly worth noting and should be made good use of for the early sowing of **tomatoes** and **cucumber** under glass.

During planting time **vines, fruit trees** and **hedges** should be pruned, preferably selecting Flower and Fruit days. Avoid unfavourable days (- - - - -).

Planet (naked eye) visibility
Evening: Venus
All night: Saturn
Morning: Mercury, Jupiter (from end of month)

March 2009

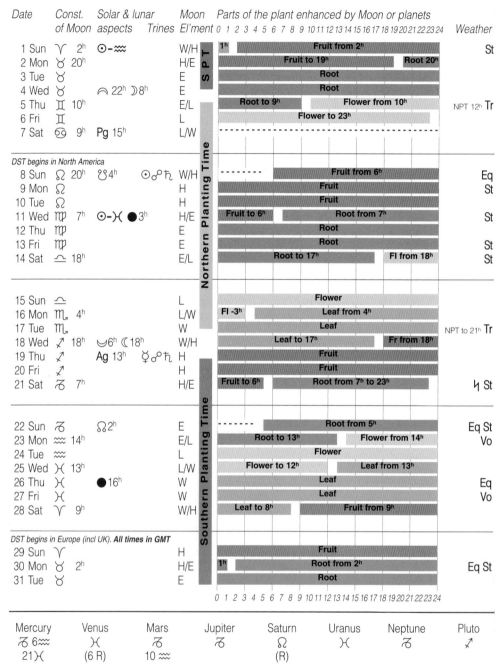

Date	Const. of Moon	Solar & lunar aspects	Trines	Moon El'ment	Parts of the plant enhanced by Moon or planets	Weather
1 Sun	♈ 2ʰ	☉-♒		W/H	1ʰ Fruit from 2ʰ	St
2 Mon	♉ 20ʰ			H/E	Fruit to 19ʰ · Root 20ʰ	
3 Tue	♉			E	Root	
4 Wed	♉		⌒ 22ʰ ☽8ʰ	E	Root	
5 Thu	♊ 10ʰ			E/L	Root to 9ʰ · Flower from 10ʰ	NPT 12ʰ Tr
6 Fri	♊			L	Flower to 23ʰ	
7 Sat	♋ 9ʰ	Pg 15ʰ		L/W		

DST begins in North America

8 Sun	♌ 20ʰ	☌4ʰ	☉♂♄	W/H	Fruit from 6ʰ	Eq St
9 Mon	♌			H	Fruit	St
10 Tue	♌			H	Fruit	
11 Wed	♍ 7ʰ	☉-♓ ●3ʰ		H/E	Fruit to 6ʰ · Root from 7ʰ	St
12 Thu	♍			E	Root	
13 Fri	♍			E	Root	St
14 Sat	♎ 18ʰ			E/L	Root to 17ʰ · Fl from 18ʰ	St

15 Sun	♎			L	Flower	
16 Mon	♏ 4ʰ			L/W	Fl -3ʰ · Leaf from 4ʰ	
17 Tue	♏			W	Leaf	NPT to 21ʰ Tr
18 Wed	♐ 18ʰ	☍6ʰ ☾18ʰ		W/H	Leaf to 17ʰ · Fr from 18ʰ	
19 Thu	♐	Ag 13ʰ	☿♂♄	H	Fruit	
20 Fri	♐			H	Fruit	
21 Sat	♑ 7ʰ			H/E	Fruit to 6ʰ · Root from 7ʰ to 23ʰ	♄ St

22 Sun	♑	♌2ʰ		E	Root from 5ʰ	Eq St
23 Mon	♒ 14ʰ			E/L	Root to 13ʰ · Flower from 14ʰ	Vo
24 Tue	♒			L	Flower	
25 Wed	♓ 13ʰ			L/W	Flower to 12ʰ · Leaf from 13ʰ	
26 Thu	♓	●16ʰ		W	Leaf	Eq
27 Fri	♓			W	Leaf	Vo
28 Sat	♈ 9ʰ			W/H	Leaf to 8ʰ · Fruit from 9ʰ	

*DST begins in Europe (incl UK). **All times in GMT***

29 Sun	♈			H	Fruit	
30 Mon	♉ 2ʰ			H/E	1ʰ Root from 2ʰ	Eq St
31 Tue	♉			E	Root	

Mercury	Venus	Mars	Jupiter	Saturn	Uranus	Neptune	Pluto
♑ 6♒	♓	♑	♑	♌	♓	♑	♐
21♓	(6 R)	10♒		(R)			

NB: All zodiac symbols refer to astronomical constellations, not astrological signs (see p.9)

Planetary aspects
(Bold = visible to naked eye)

1	
2	☿☌♂ 3ʰ
3	
4	
5	☿☌♅ 10ʰ
6	
7	
8	♂☌♅ 13ʰ ☽☍♃ 13ʰ ☉☍♄ 20ʰ
9	☽☌♅ 7ʰ ☽☍♂ 8ʰ ☽☍☿ 18ʰ
10	**☽☌♄ 22ʰ**
11	☾☍☊ 6ʰ
12	☾☍♀ 21ʰ
13	☉☌☊ 1ʰ
14	
15	
16	
17	
18	
19	☿☍♄ 4ʰ
20	
21	
22	☿☌☊ 5ʰ **☾☌♃ 21ʰ**
23	☾☌♅ 12ʰ
24	☾☌♂ 11ʰ
25	☾☍♄ 5ʰ ☾☌☊ 17ʰ
26	☾☌☿ 7ʰ **☽☌♀ 19ʰ**
27	☉☌♀ 19ʰ
28	
29	☿☌♀ 3ʰ
30	
31	☉☌☿ 4ʰ

Planet (naked eye) visibility
Evening: Venus to March 27
All night: Saturn
Morning: Jupiter, Venus from March 22(!)

March 2009

The first ten days will remain very wintry with four planets in the mid-winter constellation of Capricorn. Mercury moves into the Light region of Aquarius on March 6 and Mars follows on March 10, while Jupiter remains there for a further two months. Saturn continues to bring some warmth from Leo and should moderate the wintry effects. Venus stays in the Water region of Pisces where it is joined by Mercury on March 21.

Northern Planting Time
March 5 12ʰ to March 17 21ʰ.
Southern Planting Time
Feb 20 to March 4 19ʰ
and March 19 19ʰ to March 31 23ʰ.

March 19 is a particularly good day for cutting **scions** from fruit trees for later grafting or taking cuttings. These latter should be planted in the next planting period, for example on April 5.

Times to pick the preparation plants
Pick **dandelions** in the morning on Flower days as soon as they are open and while the centre of the flowers are still tightly packed.

Pick **yarrow** on Fruit days when the Sun is in Lion (around middle of August)

Pick **chamomile** on Flower days just before midsummer. If they are harvested too late, seeds will begin to form and there are often grubs in the hollow heads.

Collect **stinging nettles** when the first flowers are opening, usually around midsummer. Harvest the whole plants without roots on Flower days.

Pick **valerian** on Flower days around midsummer.

All the flowers (except valerian) should be laid out on paper and dried in the shade.

Collect **oak bark** on Root days. The pithy material below the bark should not be used.

April 2009

All times in GMT

Date	Const. of Moon	Solar & lunar aspects	Trines	Moon El'ment		Parts of the plant enhanced by Moon or planets (0–24h)	Weather
1 Wed	♊ 15h	☉-♓ ⌒ 3h		E/L		Root to 14h	
2 Thu	♊	Pg 2h ☽15h		L	Northern Planting Time	Flower from 17h	NPT from 17h
3 Fri	♋ 15h			L/W		Flower to 14h; Leaf from 15h	Tr Vo
4 Sat	♋	☋7h		W		Lf to 4h; Fruit from 10h	
DST ends in Australia and New Zealand							
5 Sun	♌ 3h		♂☍♄	W/H		Fruit to 23h	St
6 Mon	♌		☿☊	H			Eq
7 Tue	♍ 16h			H/E		Root from 16h	
8 Wed	♍			E		Root	
9 Thu	♍	●15h		E		Root to 23h	ϟ
10 Fri	♍	*Good Friday*		E			
11 Sat	♎ 3h		△	E/L			St
12 Sun	♏ 13h	*Easter*		L/W		Flower from 0h to 12h; Leaf from 13h	
13 Mon	♏			W		Leaf	Tr St
14 Tue	♏	◡13h		W		Leaf	NPT to 9h ϟ
15 Wed	♐ 2h			W/H	Southern Planting Time	1h; Fruit from 2h	
16 Thu	♐	Ag 9h		H		Fruit	
17 Fri	♑ 15h	☾14h		H/E		Fruit	
18 Sat	♑	☊5h	△	E		-2h; Root from 8h	
19 Sun	♒ 23h	☉-♈		E/L		Root to 22h; 23	Vo
20 Mon	♒			L		Flower	ϟ St
21 Tue	♓ 22h			L/W		Flower to 21h; 22h	
22 Wed	♓		♀●	W		Leaf to 8h; 16h-19h; Fr 20h	
23 Thu	♓		△	W		Fruit to 11h; Leaf from 12h	
24 Fri	♈ 18h			W/H		Leaf to 17h; Fr from 18h	
25 Sat	♈	●3h		H		Fruit	Tr Eq St
26 Sun	♉ 10h			H/E		Fruit to 9h; Root from 10h	Vo St
27 Mon	♉			E		Root to 17h	Vo St
28 Tue	♊ 21h	⌒ 9h Pg 6h		E/L	NPT	Fl 21h	NPT 23h Tr
29 Wed	♊			L		Flower	
30 Thu	♋ 20h			L/W		Flower to 19h; Lf 20h	Eq ϟ

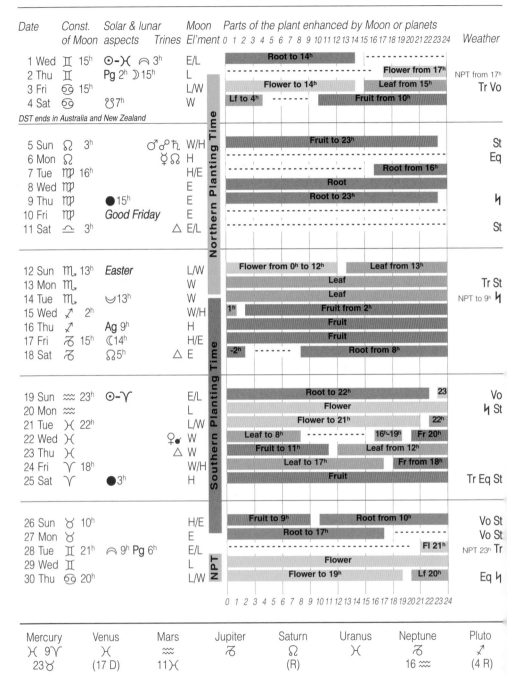

Mercury	Venus	Mars	Jupiter	Saturn	Uranus	Neptune	Pluto
♓ 9♈ 23♉	♓ (17 D)	♒ 11♓	♑	♌ (R)	♓	♑ 16 ♒	♐ (4 R)

NB: All zodiac symbols refer to astronomical constellations, not astrological signs *(see p.9)*

Planetary aspects

*(**Bold** = visible to naked eye)*

1

2

3

4

5   $\sigma \!\!\!\!\!^\circ\!\!\!\!\raisebox{0pt}{}\, \hbar$ 1ʰ $\mathbb{D} \!\!\!\!\!^\circ\!\!\!\!\, \, 4$ 6ʰ $\mathbb{D} \!\!\!\!\!^\circ\!\!\!\!\, \, \Psi$ 16ʰ

6

7   $\mathbb{D}\sigma \hbar$ 3ʰ $\mathbb{D} \!\!\!\!\!^\circ\!\!\!\!\, \, \sigma$ 6ʰ $\mathbb{D} \!\!\!\!\!^\circ\!\!\!\!\, \hat{\odot}$ 17ʰ

8   $\mathbb{D} \!\!\!\!\!^\circ\!\!\!\!\, \, \varphi$ 5ʰ

9

10   $\mathbb{C} \!\!\!\!\!^\circ\!\!\!\!\, \, \varphi$ 13ʰ

11   $\varphi \triangle \mathrm{P}$ 6ʰ

12

13

14

15   $\sigma \sigma \hat{\odot}$ 10ʰ

16

17

18   $\varphi \triangle \hbar$ 1ʰ

19   $\mathbb{C} \sigma 4$ 15ʰ $\mathbb{C} \sigma \Psi$ 22ʰ

20

21   $\mathbb{C} \!\!\!\!\!^\circ\!\!\!\!\, \, \hbar$ 11ʰ $\varphi \sigma \sigma$ 23ʰ

22   $\mathbb{C} \sigma \hat{\odot}$ 5ʰ $\mathbb{C} \bullet \varphi$ 13ʰ $\mathbb{C} \sigma \sigma$ 14ʰ

23   $\odot \triangle \mathrm{P}$ 5ʰ

24

25

26   $\mathbb{D} \sigma \varphi$ 16ʰ

27

28

29

30

Planet (naked eye) visibility
Evening: Mercury
All night: Saturn
Morning: Venus, Jupiter

April 2009

At the beginning of the month Mercury and Venus are in the Watery region of Pisces. They are joined on April 11 by Mars. On April 9, Mercury moves into the Warmth constellation of Aries and stays there until April 22. Now the soil is warming up and nights are warmer too. On April 16 Neptune moves into Aquarius and this will reduce the Warmth effect, but should enhance Light.

Northern Planting Time
April 2 17ʰ to April 14 9ʰ
and April 28 23ʰ to May 9.
Southern Planting Time
April 14 18ʰ to April 27 17ʰ.

The best period for **grafting** is especially between April 15 and 17, or if that is not possible April 23 to 25.

April 25 is also a very good day to plant **potatoes for use as seed** in 2010. For a **potato crop,** root days should be chosen.

Biodynamic preparations: between April 7 5ʰ and 8 7ʰ cut birch, fill with yarrow and hang.

Stag's bladder and birch-yarrow preparation.

May 2009

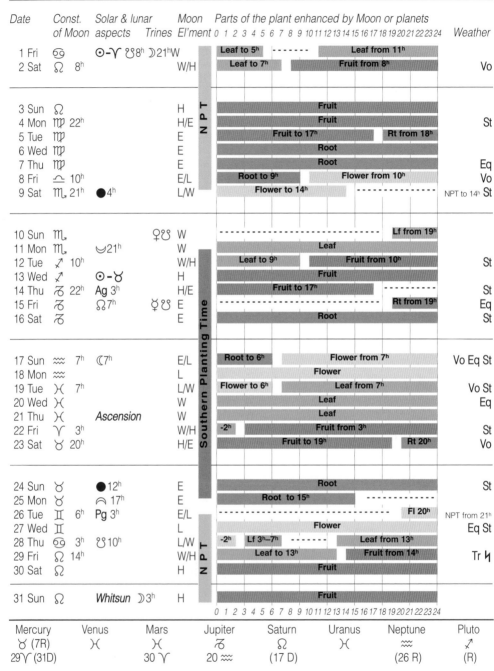

Date	Const. of Moon	Solar & lunar aspects	Moon Trines	El'ment	Parts of the plant enhanced by Moon or planets 0 1 2 3 4 5 6 7 8 9 10 11 12 13 14 15 16 17 18 19 20 21 22 23 24	Weather
1 Fri	♋	☉-♈ ☍8ʰ ☽21ʰW			Leaf to 5ʰ - - - - - - Leaf from 11ʰ	
2 Sat	♌ 8ʰ			W/H	Leaf to 7ʰ Fruit from 8ʰ	Vo
3 Sun	♌			H	Fruit	
4 Mon	♍ 22ʰ			H/E	Fruit	St
5 Tue	♍			E	Fruit to 17ʰ Rt from 18ʰ	
6 Wed	♍			E	Root	
7 Thu	♍			E	Root	Eq
8 Fri	♎ 10ʰ			E/L	Root to 9ʰ Flower from 10ʰ	Vo
9 Sat	♏ 21ʰ	●4ʰ		L/W	Flower to 14ʰ - - - - - - - - -	NPT to 14ʰ St
10 Sun	♏		♀☍	W	- - - - - - - - - - - - - - - - Lf from 19ʰ	
11 Mon	♏	◡21ʰ		W	Leaf	
12 Tue	♐ 10ʰ			W/H	Leaf to 9ʰ Fruit from 10ʰ	St
13 Wed	♐	☉-♉		H	Fruit	
14 Thu	♑ 22ʰ	Ag 3ʰ		H/E	Fruit to 17ʰ - - - - - - -	St
15 Fri	♑	♌☌7ʰ	☿☍	E	- - - - - - - - - - - - - - - - Rt from 19ʰ	Eq
16 Sat	♑			E	Root	St
17 Sun	♒ 7ʰ	☾7ʰ		E/L	Root to 6ʰ Flower from 7ʰ	Vo Eq St
18 Mon	♒			L	Flower	
19 Tue	♓ 7ʰ			L/W	Flower to 6ʰ Leaf from 7ʰ	Vo St
20 Wed	♓			W	Leaf	Eq
21 Thu	♓	Ascension		W	Leaf	
22 Fri	♈ 3ʰ			W/H	-2ʰ Fruit from 3ʰ	St
23 Sat	♉ 20ʰ			H/E	Fruit to 19ʰ Rt 20ʰ	Vo
24 Sun	♉	●12ʰ		E	Root	St
25 Mon	♉	◠17ʰ		E	Root to 15ʰ - - - - - - -	
26 Tue	♊ 6ʰ	Pg 3ʰ		E/L	- - - - - - - - - - - - - - - - Fl 20ʰ	NPT from 21ʰ
27 Wed	♊			L	Flower	Eq St
28 Thu	♋ 3ʰ	☍10ʰ		L/W	-2ʰ Lf 3ʰ–7ʰ - - - - - - Leaf from 13ʰ	
29 Fri	♌ 14ʰ			W/H	Leaf to 13ʰ Fruit from 14ʰ	Tr ♄
30 Sat	♌			H	Fruit	
31 Sun	♌	Whitsun ☽3ʰ		H	Fruit	

0 1 2 3 4 5 6 7 8 9 10 11 12 13 14 15 16 17 18 19 20 21 22 23 24

Mercury	Venus	Mars	Jupiter	Saturn	Uranus	Neptune	Pluto
♉ (7R)	♓	♓	♑	♌	♓	♒	♐
29♈ (31D)		30 ♈	20 ♒	(17 D)		(26 R)	(R)

24   *NB: All zodiac symbols refer to astronomical constellations, not astrological signs (see p.9)*

Planetary aspects

*(**Bold** = visible to naked eye)*

1

2 ☽☍♃ 18^h ☽☍♆ 22^h

3

4 **☽☌♄** 7^h

5 ☽☍♅ 2^h ☉△♄ 9^h ☽☍♀ 18^h

6 ☽☍♂ 5^h

7

8

9

10 ☾☍☿ 5^h

11

12

13

14

15

16

17 **☾☌♃** 6^h ☾☌♆ 7^h

18 ☉☌☿ 10^h ☾☍♄ 19^h

19 ☾☌♅ 16^h

20

21 **☾☌♀** 3^h ☾☌♂ 15^h

22

23 ☾☌☿ 22^h

24

25

26

27 ♃☌♆ 22^h

28

29

30 ☽☍♆ 4^h ☽☍♃ 4^h

31 **☽☌♄** 13^h

Planet (naked eye) visibility

Evening:

All night: Saturn

Morning: Venus, Jupiter

May 2009

From the last week in April Mercury is in the Earth constellation of Taurus and willremain ther for most of May. This means the nights will be cooler again. Venus and Mars are still in the Watery region of Pisces. In the middle of the month Mars moves into Aries but unlike Mercury has little influence on warmth. Jupiter moves into the Light region of Aquarius on May 20. This enhances its own qualities and should lead to plenty of light in the coming two months. Saturn and Pluto remain in Warmth constellations. Planetary positioning therefore suggests positive conditions for plant growth.

Northern Planting Time
April 29 to May 9 14^h
and May 26 21^h to June 7.
Southern Planting Time
May 11 21^h to May 25 11^h.

Control **moths** from May 22 3^h until 23 20^h.

Control **varroa and chitinous insects** on May 24 and 25.

June 2009

Date	Const. of Moon	Solar & lunar aspects	Moon Trines	El'ment	Parts of the plant enhanced by Moon or planets	Weather

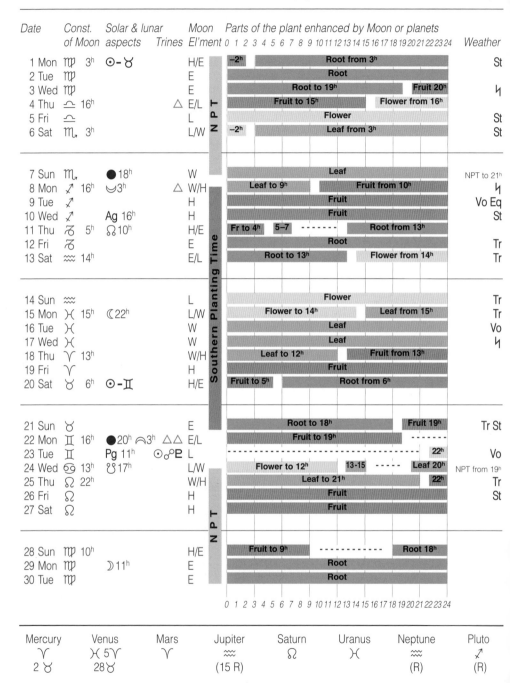

Date	Const. of Moon	Solar & lunar aspects	Trines	El'ment	Parts of the plant (0–24)	Weather
1 Mon	♍ 3ʰ	☉-♉		H/E	–2ʰ / Root from 3ʰ	St
2 Tue	♍			E	Root	
3 Wed	♍			E	Root to 19ʰ / Fruit 20ʰ	♄
4 Thu	♎ 16ʰ		△	E/L	Fruit to 15ʰ / Flower from 16ʰ	
5 Fri	♎			L	Flower	St
6 Sat	♏ 3ʰ			L/W	–2ʰ / Leaf from 3ʰ	St
7 Sun	♏	● 18ʰ		W	Leaf	NPT to 21ʰ
8 Mon	♐ 16ʰ	☋ 3ʰ	△	W/H	Leaf to 9ʰ / Fruit from 10ʰ	♄
9 Tue	♐			H	Fruit	Vo Eq
10 Wed	♐	Ag 16ʰ		H	Fruit	St
11 Thu	♑ 5ʰ	♋ 10ʰ		H/E	Fr to 4ʰ / 5–7 / Root from 13ʰ	
12 Fri	♑			E	Root	Tr
13 Sat	♒ 14ʰ			E/L	Root to 13ʰ / Flower from 14ʰ	Tr
14 Sun	♒			L	Flower	Tr
15 Mon	♓ 15ʰ	☾ 22ʰ		L/W	Flower to 14ʰ / Leaf from 15ʰ	Tr
16 Tue	♓			W	Leaf	Vo
17 Wed	♓			W	Leaf	♄
18 Thu	♈ 13ʰ			W/H	Leaf to 12ʰ / Fruit from 13ʰ	
19 Fri	♈			H	Fruit	
20 Sat	♉ 6ʰ	☉-♊		H/E	Fruit to 5ʰ / Root from 6ʰ	
21 Sun	♉			E	Root to 18ʰ / Fruit 19ʰ	Tr St
22 Mon	♊ 16ʰ	● 20ʰ ☊ 3ʰ	△△	E/L	Fruit to 19ʰ	
23 Tue	♊	Pg 11ʰ	☉☌♇	L	22ʰ	Vo
24 Wed	♋ 13ʰ	☍ 17ʰ		L/W	Flower to 12ʰ / 13–15 / Leaf 20ʰ	NPT from 19ʰ
25 Thu	♌ 22ʰ			W/H	Leaf to 21ʰ / 22ʰ	Tr
26 Fri	♌			H	Fruit	St
27 Sat	♌			H	Fruit	
28 Sun	♍ 10ʰ			H/E	Fruit to 9ʰ / Root 18ʰ	
29 Mon	♍	☽ 11ʰ		E	Root	
30 Tue	♍			E	Root	

Mercury	Venus	Mars	Jupiter	Saturn	Uranus	Neptune	Pluto
♈	♓ 5♈	♈	♒	♌	♓	♒	♐
2 ♉	28 ♉		(15 R)			(R)	(R)

NB: All zodiac symbols refer to astronomical constellations, not astrological signs (see p.9)

Planetary aspects

(Bold = visible to naked eye)

1	☽ ☌ ⚷ 9ʰ
2	
3	☽ ☌ ♀ 18ʰ
4	☽ ☌ ♂ 3ʰ ♂ △ ♇ 4ʰ
5	☽ ☌ ☿ 21ʰ
6	
7	
8	♀ △ ♇ 22ʰ
9	
10	
11	
12	
13	☾ ☌ ♆ 14ʰ ☾ ☌ ♃ 16ʰ
14	
15	☾ ☌ ♄ 4ʰ
16	☾ ☌ ⚷ 1ʰ
17	☉ △ ♆ 11ʰ
18	☉ △ ♃ 2ʰ
19	☾ ☌ ♀ 13ʰ ☾ ☌ ♂ 14ʰ
20	
21	☾ ☌ ☿ 7ʰ ♀ ☌ ♂ 13ʰ
22	♀ △ ♄ 7ʰ ♂ △ ♄ 15ʰ
23	
24	
25	
26	☽ ☌ ♆ 12ʰ ☽ ☌ ♃ 12ʰ
27	☽ ☌ ♄ 22ʰ
28	☽ ☌ ⚷ 15ʰ
29	
30	

Planet (naked eye) visibility
Evening: Saturn
All night:
Morning: Venus, Jupiter, Mars (from mid-month)

June 2009

Mercury is in Taurus. Venus is in Aries from June 5 to 27 and mediates its beneficial influence to the earth. Mars is still in Aries. In June we can therefore expect warm, sunny weather and good growing conditions. This is supported by four Warmth trines. It is questionable whether Uranus on its own in Pisces will be able to induce moisture and it could be that we are in need of rain.

Northern Planting Time
May 26 to June 7 21ʰ and June 24 19ʰ to July 5.
Southern Planting Time
June 8 11ʰ to June 21 21ʰ.

Control **mole crickets** from June 6 3ʰ to June 8 15ʰ. Burn fly papers in the cowshed on Flower days.

To control **ants** in the house burn them when the Moon is in Leo.

Control **grasshoppers** June 22 16ʰ to June 24 11ʰ.

Haymaking best from June 5 to 27. Cut on Flower days.

Fungal problems
The function of fungus in nature is to break down dying organic materials. It appears amongst our crops when unripe manure compost or uncomposted animal by-products such as horn and bone meal are used but also when seeds are harvested during unfavourable constellations. 'When moon forces are working too strongly on the earth ...' (Steiner, *Agriculture Course)* tea can be made from horsetail *(Equisetum arvense)* and sprayed on to the soil where effected plants are growing. This draws the fungal level back down into the ground where it belongs.

The plants can be strengthened by spraying stinging nettle tea on the leaves. This will promote good assimilation, stimulate the flow of sap and cause fungus diseases to disappear.

June

July 2009

All times in GMT

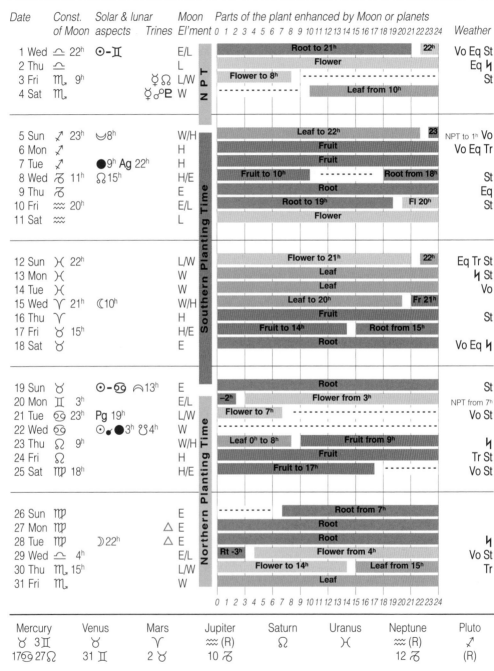

Date	Const. of Moon	Solar & lunar aspects / Trines	Moon El'ment	Parts of the plant enhanced by Moon or planets	Weather
1 Wed	♎ 22ʰ	☉-♊	E/L	Root to 21ʰ · 22ʰ	Vo Eq St
2 Thu	♎		L	Flower	Eq ♄
3 Fri	♏ 9ʰ	☿ ♌	L/W	Flower to 8ʰ · · · · · ·	St
4 Sat	♏	☿ ♂ ♇	W	· · · · · · Leaf from 10ʰ	
				NPT	
5 Sun	♐ 23ʰ	☽8ʰ	W/H	Leaf to 22ʰ · 23	NPT to 1ʰ Vo
6 Mon	♐		H	Fruit	Vo Eq Tr
7 Tue	♐	●9ʰ Ag 22ʰ	H	Fruit	
8 Wed	♑ 11ʰ	♌15ʰ	H/E	Fruit to 10ʰ · · · · · · Root from 18ʰ	St
9 Thu	♑		E	Root	Eq
10 Fri	♒ 20ʰ		E/L	Root to 19ʰ · Fl 20ʰ	St
11 Sat	♒		L	Flower	
				Southern Planting Time	
12 Sun	♓ 22ʰ		L/W	Flower to 21ʰ · 22ʰ	Eq Tr St
13 Mon	♓		W	Leaf	♄ St
14 Tue	♓		W	Leaf	Vo
15 Wed	♈ 21ʰ	☽10ʰ	W/H	Leaf to 20ʰ · Fr 21ʰ	
16 Thu	♈		H	Fruit	St
17 Fri	♉ 15ʰ		H/E	Fruit to 14ʰ · Root from 15ʰ	
18 Sat	♉		E	Root	Vo Eq ♄
				Northern Planting Time	
19 Sun	♉	☉-♋ ⌒13ʰ	E	Root	St
20 Mon	♊ 3ʰ		E/L	-2ʰ · Flower from 3ʰ	NPT from 7ʰ
21 Tue	♋ 23ʰ	Pg 19ʰ	L/W	Flower to 7ʰ · · · · · ·	Vo St
22 Wed	♋	☉ ♂ ●3ʰ ♌4ʰ	W	· · · · · ·	
23 Thu	♌ 9ʰ		W/H	Leaf 0ʰ to 8ʰ · Fruit from 9ʰ	♄
24 Fri	♌		H	Fruit	Tr St
25 Sat	♍ 18ʰ		H/E	Fruit to 17ʰ · · · · · ·	Vo St
				Northern Planting Time	
26 Sun	♍		E	· · · · · · Root from 7ʰ	St
27 Mon	♍	△	E	Root	
28 Tue	♍	☽22ʰ △	E	Root	♄
29 Wed	♎ 4ʰ		E/L	Rt -3ʰ · Flower from 4ʰ	Vo St
30 Thu	♏ 15ʰ		L/W	Flower to 14ʰ · Leaf from 15ʰ	Tr
31 Fri	♏		W	Leaf	

0 1 2 3 4 5 6 7 8 9 10 11 12 13 14 15 16 17 18 19 20 21 22 23 24

Mercury	Venus	Mars	Jupiter	Saturn	Uranus	Neptune	Pluto
♉ 3♊	♉	♈	♒ (R)	♌	♓	♒ (R)	♐
17♋ 27♌	31 ♊	2 ♉	10 ♑			12 ♑	(R)

28   *NB: All zodiac symbols refer to astronomical constellations, not astrological signs (see p.9)*

Planetary aspects

*(**Bold** = visible to naked eye)*

1 ☿△♅ 21ʰ
2 ☿△♃ 2ʰ
3 ☽☍♂ 2ʰ ☽☍♀ 10ʰ
4 ☿☌♇ 16ʰ

5
6 ☽☌☿ 14ʰ
7
8
9
10 ♃☌♅ 7ʰ ☽☌♃ 20ʰ ☽☌♅ 20ʰ
11

12 ☽☍♄ 14ʰ
13 ☽☌☋ 8ʰ
14 ☉☌☿ 2ʰ
15
16 ☿△☋ 8ʰ
17
18 ☽☌♂ 10ʰ

19 ☉△☋ 0ʰ ☽☌♀ 4ʰ
20
21
22 ☽☌☿ 19ʰ
23 ☽☍♃ 19ʰ ☽☍♅ 20ʰ
24
25 ☽☌♄ 10ʰ ☽☍☋ 23ʰ

26
27 ♀△♃ 4ʰ
28 ♀△♅ 6ʰ
29
30 ☿☍♃ 11ʰ
31 ☿☍♅ 8ʰ ☽☍♂ 23ʰ

Planet (naked eye) visibility
Evening: Saturn
All night: Jupiter
Morning: Venus, Mars

July 2009

Mercury passes rapidly through the summer constellations in order to arrive on July 27 in Leo from where its influence will help to ripen the grain. Venus and Mars remain throughout July in the Earth constellation of Taurus. Jupiter leaves its favourite constellation Aquarius and returns to Capricorn for the rest of the year accompanying Neptune on its slow journey through this Earth constellation. The winter forces of Capricorn will increase the likelihood of cool nights.

Northern Planting Time
June 24 to July 5 1ʰ and July 20 7ʰ to Aug 1.
Southern Planting Time
July 5 8ʰ to July 19 11ʰ.

After July 16 **snails** and **slugs** will appear among the lettuces. They do not like horn silica. This should be sprayed widely across the ground among the leafy plants early in the morning of July 16.
Control **grasshoppers** July 20 3ʰ to 21 22ʰ.

Supplementary feeding of bees
Many years of experience has demonstrated the health value of supplementing winter feeding with herb teas. Yarrow, chamomile and valerian flowers are steeped in boiling water for about fifteen minutes. Stinging nettle, horsetail and oak bark put in cold water and then boiled for ten minutes. All are sieved and mixed together. Three grams of each dried herb is enough to produce 100 litres of liquid. This treatment is particularly important in years when the final honey of year comes from honeydew.

July

August 2009

All times in GMT

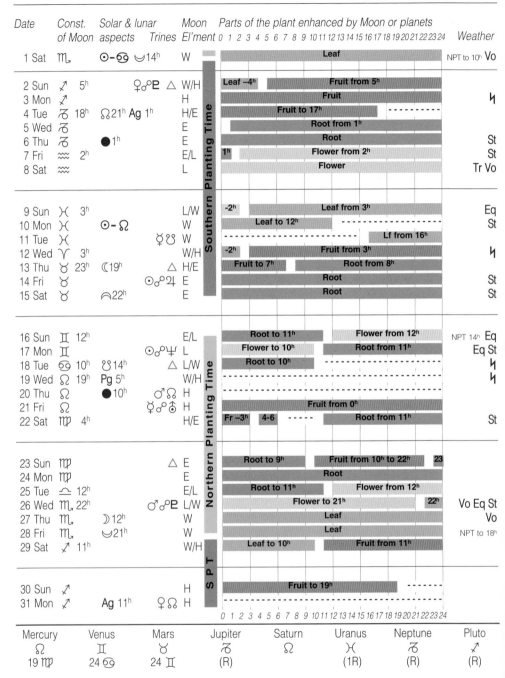

Date	Const. of Moon	Solar & lunar aspects	Moon Trines	El'ment	Parts of the plant enhanced by Moon or planets (0–24)	Weather
1 Sat	♏	☉-♋ ☽14ʰ		W	Leaf	NPT to 10ʰ Vo
2 Sun	♐ 5ʰ	♀☌♇ △	W/H		Leaf –4ʰ / Fruit from 5ʰ	
3 Mon	♐		H		Fruit	♄
4 Tue	♑ 18ʰ	♌21ʰ Ag 1ʰ	H/E		Fruit to 17ʰ	
5 Wed	♑		E		Root from 1ʰ	
6 Thu	♑	●1ʰ	E		Root	St
7 Fri	♒ 2ʰ		E/L		1ʰ Flower from 2ʰ	St
8 Sat	♒		L		Flower	Tr Vo
9 Sun	♓ 3ʰ		L/W		–2ʰ Leaf from 3ʰ	Eq
10 Mon	♓	☉-♌	W		Leaf to 12ʰ	St
11 Tue	♓	☿♋	W		Lf from 16ʰ	
12 Wed	♈ 3ʰ		W/H		–2ʰ Fruit from 3ʰ	♄
13 Thu	♉ 23ʰ	☽19ʰ △	H/E		Fruit to 7ʰ / Root from 8ʰ	
14 Fri	♉	☉☌♃	E		Root	St
15 Sat	♉	☌22ʰ	E		Root	St
16 Sun	♊ 12ʰ		E/L		Root to 11ʰ / Flower from 12ʰ	NPT 14ʰ Eq
17 Mon	♊	☉☌♆	L		Flower to 10ʰ / Root from 11ʰ	Eq St
18 Tue	♋ 10ʰ	☍14ʰ △	L/W		Root to 10ʰ	♄
19 Wed	♌ 19ʰ	Pg 5ʰ	W/H			♄
20 Thu	♌	●10ʰ	H		♂♌	
21 Fri	♌	☿☌♁	H		Fruit from 0ʰ	
22 Sat	♍ 4ʰ		H/E		Fr –3ʰ 4-6 Root from 11ʰ	St
23 Sun	♍	△	E		Root to 9ʰ / Fruit from 10ʰ to 22ʰ 23	
24 Mon	♍		E		Root	
25 Tue	♎ 12ʰ		E/L		Root to 11ʰ / Flower from 12ʰ	
26 Wed	♏ 22ʰ	♂☌♇	L/W		Flower to 21ʰ 22ʰ	Vo Eq St
27 Thu	♏	☽12ʰ	W		Leaf	Vo
28 Fri	♏	☽21ʰ	W		Leaf	NPT to 18ʰ
29 Sat	♐ 11ʰ		W/H		Leaf to 10ʰ / Fruit from 11ʰ	
30 Sun	♐		H		Fruit to 19ʰ	
31 Mon	♐	Ag 11ʰ	♀♌ H			

Southern Planting Time (2 Sun – 8 Sat)
Northern Planting Time (16 Sun – 22 Sat, 23 Sun – 29 Sat)
SPT (30 Sun – 31 Mon)

Mercury	Venus	Mars	Jupiter	Saturn	Uranus	Neptune	Pluto
♌	♊	♉	♑	♌	♓	♑	♐
19 ♍	24 ♋	24 ♊	(R)		(1R)	(R)	(R)

30

NB: All zodiac symbols refer to astronomical constellations, not astrological signs (see p.9)

Planetary aspects

*(**Bold** = visible to naked eye)*

1
2 ♀☌♇ 3ʰ ☽☍♀ 11ʰ ☿△♇ 14ʰ
3
4
5
6 **☾☌♃ 20ʰ**
7 **☾☌♆ 0ʰ**
8 **☾☍☿ 2ʰ**

9 **☾☍♄ 1ʰ ☾☌☊ 13ʰ**
10
11
12
13 ♂△♃ 17ʰ
14 ☉☍♃ 18ʰ
15

16 **☾☌♂ 3ʰ**
17 ☿☌♄ 15ʰ ☉☍♆ 21ʰ **☾☌♀ 21ʰ**
18 ♂△♆ 2ʰ
19
20 **☾☍♃ 0ʰ ☾☍♆ 6ʰ**
21 ☿☍☊ 12ʰ
22 ☽☌♄ 1ʰ ☽☍☊ 8ʰ ☽☌☿ 10ʰ

23 ♀△☊ 0ʰ
24
25
26
27
28
29 ☽☍♂ 20ʰ

30
31

Planet (naked eye) visibility

Evening: Saturn (to mid-month)
All night: Jupiter
Morning: Venus, Mars

August 2009

With Mercury in Leo and Venus in Gemini it is a good time for **harvesting cereals** and **soft fruit.** Mercury moves into Virgo on Aug 19 bringing the possibility of the first autumn mist. On Aug 21 Venus enters Cancer to once more delight the **slugs,** but they do not like horn silica.

Northern Planting Time
July 20 to Aug 1 10ʰ
and Aug 16 14ʰ to Aug 28 18ʰ.
Southern Planting Time
Aug 1 23ʰ to Aug 15 17ʰ
and Aug 29 1ʰ to Sep 11.

The harvest of **grains for seed** should take place on Fruit days. Especially favourable are August 2 and 3. As soon as harvest is complete, sow green manure catch crops like lupins, cornflowers, phacelia, mustard or linseed.

The best days for **early harvesting of fruit** is on Flower and Fruit days outside the Planting Time.

Control **ants** in houses Aug 19 19ʰ to Aug 21 23ʰ.

Aug

September 2009

All times in GMT

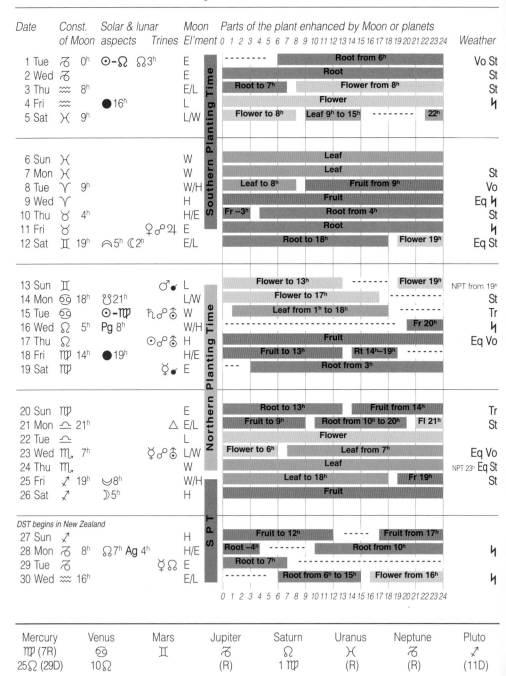

Date	Const. of Moon	Solar & lunar aspects	Trines	Moon El'ment	Parts of the plant enhanced by Moon or planets	Weather

Southern Planting Time

1 Tue	♑ 0ʰ	☉-♌ ☊3ʰ		E	Root from 6ʰ	Vo St
2 Wed	♑			E	Root	St
3 Thu	♒ 8ʰ			E/L	Root to 7ʰ — Flower from 8ʰ	St
4 Fri	♒	●16ʰ		L	Flower	♄
5 Sat	♓ 9ʰ			L/W	Flower to 8ʰ — Leaf 9ʰ to 15ʰ — 22ʰ	

6 Sun	♓			W	Leaf	
7 Mon	♓			W	Leaf	St
8 Tue	♈ 9ʰ			W/H	Leaf to 8ʰ — Fruit from 9ʰ	Vo
9 Wed	♈			H	Fruit	Eq ♄
10 Thu	♉ 4ʰ			H/E	Fr –3ʰ — Root from 4ʰ	St
11 Fri	♉	♀☌♃		E	Root	♄
12 Sat	♊ 19ʰ	◠5ʰ ☾2ʰ		E/L	Root to 18ʰ — Flower 19ʰ	Eq St

Northern Planting Time

13 Sun	♊	♂•		L	Flower to 13ʰ — Flower 19ʰ	NPT from 19ʰ
14 Mon	♋ 18ʰ	☊21ʰ		L/W	Flower to 17ʰ —	St
15 Tue	♋	☉-♍	♄☌☉	W	Leaf from 1ʰ to 18ʰ —	Tr
16 Wed	♌ 5ʰ	Pg 8ʰ		W/H	Fr 20ʰ	♄
17 Thu	♌	☉☌☉		H	Fruit	Eq Vo
18 Fri	♍ 14ʰ	●19ʰ		H/E	Fruit to 13ʰ — Rt 14ʰ–19ʰ —	
19 Sat	♍	☿•		E	Root from 3ʰ	

20 Sun	♍			E	Root to 13ʰ — Fruit from 14ʰ	Tr
21 Mon	♎ 21ʰ		△	E/L	Fruit to 9ʰ — Root from 10ʰ to 20ʰ — Fl 21ʰ	St
22 Tue	♎			L	Flower	
23 Wed	♏ 7ʰ	☿☌☉		L/W	Flower to 6ʰ — Leaf from 7ʰ	Eq Vo
24 Thu	♏			W	Leaf	NPT 23ʰ Eq St
25 Fri	♐ 19ʰ	☋8ʰ		W/H	Leaf to 18ʰ — Fr 19ʰ	St
26 Sat	♐	☽5ʰ		H	Fruit	

DST begins in New Zealand

SPT

27 Sun	♐			H	Fruit to 12ʰ — Fruit from 17ʰ	
28 Mon	♑ 8ʰ	☊7ʰ Ag 4ʰ		H/E	Root –4ʰ — Root from 10ʰ	♄
29 Tue	♑	☿♌		E	Root to 7ʰ —	
30 Wed	♒ 16ʰ			E/L	Root from 6ʰ to 15ʰ — Flower from 16ʰ	♄

0 1 2 3 4 5 6 7 8 9 10 11 12 13 14 15 16 17 18 19 20 21 22 23 24

Mercury	Venus	Mars	Jupiter	Saturn	Uranus	Neptune	Pluto
♍ (7R)	♋	♊	♑	♌	♓	♑	♐
25♌ (29D)	10♌		(R)	1♍	(R)	(R)	(11D)

NB: All zodiac symbols refer to astronomical constellations, not astrological signs (see p.9)

Planetary aspects

*(**Bold** = visible to naked eye)*

1 ☽ ☌° ♀ 18ʰ
2 **☽ ☌ ♃ 19ʰ**
3 ☽ ☌ ♆ 5ʰ
4
5 ☾ ☌° ♄ 14ʰ   ☾ ☌ ☊ 17ʰ

6 ☾ ☌° ☿ 14ʰ
7
8
9
10
11 ♀ ☌° ♃ 8ʰ
12

13 **☾ ☌ ♂ 16ʰ**
14
15 ♄ ☌ ☊ 12ʰ   ♀ ☌° ♆ 22ʰ
16 ☾ ☌° ♃ 5ʰ   ☾ ☌° ♆ 15ʰ   **☾ ☌ ♀ 16ʰ**
17 ☉ ☌° ☊ 10ʰ   ☉ ☌ ♄ 18ʰ
18 ☾ ☌° ☊ 16ʰ   ☾ ☌ ♄ 17ʰ
19 ☽ ☌ ☿ 0ʰ

20 ☉ ☌ ☿ 10ʰ
21 ♀ △ ♇ 3ʰ
22 ☿ ☌ ♄ 9ʰ
23 ☿ ☌° ☊ 16ʰ
24
25
26

27 ☽ ☌° ♂ 14ʰ
28
29 **☽ ☌ ♃ 22ʰ**
30 ☽ ☌ ♆ 12ʰ

Planet (naked eye) visibility

Evening:
All night: Jupiter
Morning: Venus, Mars

September 2009

Mercury in Virgo brings morning mists, but because Mars is in Gemini and from Sep 10 Venus is in Leo, there will be more Light qualities, too. Mercury goes back into Leo on Sep 25 and will bring about a week of warm weather. Saturn leaves Leo and will be in Virgo for the next five years. It will mediate autumnal forces.

Northern Planting Time
Sep 13 19ʰ to Sep 24 23ʰ.
Southern Planting Time
Aug 29 to Sep 11 20ʰ and Sep 25 9ʰ to Oct 9.

The best days for **harvesting fruit** occur when the Moon is in Aries or Sagittarius.

Root crops are always best when harvested on Root days as has been demonstrated through storage trials with onions, carrots, beetroot and potatoes, The last week in September is recommended.

A good time for sowing **winter cereals** is Sep 26 and 27.

Favourable times making biodynamic preparations
Sep 10 11ʰ to Sep 11 17ʰ cut maple and **dandelion** and bury in earth. Bury birch and yarrow.

Sep 22 23ʰ to Sep 23 23ʰ cut larch and **chamomile** and bury in earth.

Sep 27 6ʰ to 17ʰ cut oak and **oak bark** and bury in earth.

Sep

October 2009

All times in GMT

Date	Const. of Moon	Solar & lunar aspects	Trines	Moon El'ment	Planting time	Parts of the plant enhanced by Moon or planets (0–24)	Weather
1 Thu	♒	☉-♍		L	SPT	Flower	
2 Fri	♓ 17^{h}			L/W	SPT	Flower to 16^{h}; Leaf from 17^{h}	Eq ♄
3 Sat	♓			W	SPT	1^{h} - - - - - - - Leaf from 7^{h}	St
DST begins in Australia							
4 Sun	♓	● 6^{h}	☿☍⛢	W	SPT	Leaf	♄
5 Mon	♈ 15^{h}			W/H	SPT	Leaf to 14^{h}; Fruit from 15^{h}	
6 Tue	♈			H	SPT	Fruit	♄
7 Wed	♉ 10^{h}			H/E	SPT	Fruit to 9^{h}; Root from 10^{h}	
8 Thu	♉			E	SPT	Root	
9 Fri	♉	⌒ 10^{h}	♀☍⛢	E		Root	St
10 Sat	♊ 1^{h}		△	E/L	Northern Planting Time	Root to 11^{h}; Flower from 12^{h}	NPT 3^{h} Eq Vo
11 Sun	♊	☋ 23^{h} ☾ 9^{h}		L	Northern Planting Time	Flower to 16^{h} - - - - - - - - - - - -	St
12 Mon	♋ 0^{h}		♂●	W	Northern Planting Time	- - - - - Leaf from 4^{h} to 23^{h}	Tr
13 Tue	♌ 12^{h}	Pg 12^{h}		W/H	Northern Planting Time	- -	
14 Wed	♌			H	Northern Planting Time	Fruit from 0^{h}	Eq
15 Thu	♍ 23^{h}			H/E	Northern Planting Time	Fruit to 22^{h}; 23	Eq Vo
16 Fri	♍			E	Northern Planting Time	Root	St
17 Sat	♍		△	E	Northern Planting Time	Root	
18 Sun	♍	● 6^{h}		E	Northern Planting Time	Root	Tr Vo
19 Mon	♎ 7^{h}			E/L	Northern Planting Time	Root to 6^{h}; Flower from 7^{h}	
20 Tue	♏ 16^{h}		△	L/W	Northern Planting Time	Flower to 15^{h}; Root from 16^{h}	
21 Wed	♏			W	Northern Planting Time	-2^{h}; Leaf from 3^{h}	Tr Eq
22 Thu	♏	◡ 13^{h}		W	SPT	Leaf	NPT to 5^{h} ♄
23 Fri	♐ 3^{h}			W/H	SPT	-2^{h}; Fruit from 3^{h}	♄
24 Sat	♐		△	H	SPT	Fruit to 9^{h}; Root from 10^{h} to 23^{h}	
DST ends in Europe (incl UK)							
25 Sun	♑ 16^{h}	☊ 9^{h} Ag 23^{h}		H/E	SPT	Fr 0^{h} to 5^{h} - - - - - - - - 12^{h}-15^{h} Flower from 16^{h}	
26 Mon	♑	☽ 1^{h}		E	SPT	Fl to 4^{h}; Root from 5^{h}	
27 Tue	♑			E	SPT	Root to 23^{h}	Eq ♄
28 Wed	♒ 0^{h}			L	SPT	Flower from 0^{h} to 16^{h}; Root from 17^{h}	
29 Thu	♒		△	L	SPT	Root to 7^{h}; Flower from 8^{h} to 24^{h}	Tr St
30 Fri	♓ 1^{h}			L/W	SPT	Leaf from 1^{h}	Tr
31 Sat	♓			W	SPT	Leaf	♄
DST ends in N America							

Mercury	Venus	Mars	Jupiter	Saturn	Uranus	Neptune	Pluto
♌	♌	♊	♑	♍	♓	♑	♐
3 ♍	9 ♍	10 ♋	(13D)		(R)	(R)	

NB: All zodiac symbols refer to astronomical constellations, not astrological signs (see p.9)

Planetary aspects

(*Bold* = *visible to naked eye)*

1
2 ☽☍♀ 3^h ☽☍☿ 19^h ☽☌♅ 22^h
3 ☽☍♄ 4^h
4 ☿☍♅ 21^h
5 ♂△♅ 2^h
6
7
8 ☿☌♄ 7^h
9 ♀☍♅ 22^h
10 ☉△♃ 9^h
11
12 **☾☌♂ 1^h**
13 ☾☍♃ 10^h ♀☌♄ 11^h ☾☍♆ 21^h
14
15
16 ☾☍♅ 0^h ☾☌♄ 8^h **☾☌♀ 14^h**
17 ☉△♆ 1^h ☾☌☿ 5^h
18
19
20 ☿△♃ 22^h
21
22
23
24 ☿△♆ 17^h
25
26 ☽☍♂ 5^h
27 **☽☌♃ 7^h** ☽☌♆ 19^h
28
29 ♀△♃ 2^h
30 ☽☌♅ 5^h ☽☍♄ 18^h
31

Planet (naked eye) visibility

Evening: Jupiter

All night:

Morning: Venus, Saturn, Mars

October 2009

Venus and Mars are well placed to support the planting of cereals with Oct 1 and 2 being particularly favourable. From the second week and for the rest of the month Earth and Water qualities dominate. Five planets are in Earth constellations and are supported by five Earth trines. It is important to ensure the autumn cultivations are completed by this time.

Northern Planting Time

Oct 10 3^h to Oct 22 5^h.

Southern Planting Time

Sep 25 to Oct 9 5^h and Oct 22 17^h to Nov 5.

The best times for harvesting **storage fruit** are Oct 6, 10, 23, 24, 28 and 29. All harvested areas should be supplied with compost, treated with barrel preparation and dug over for winter.

Making rye bread

Wheat, barley, oats, maize, rice and millet can be milled into flour and made into bread using yeast or 'back ferment'. With rye however another procedure is necessary. Finnish researchers have discovered that predisposition to cancer can disappear by eating rye bread. A detailed recipe for making rye bread is on page 53.

Rye needs a room temperature of 28°C (82°F) and I was amazed how during a heat wave my bread dough completed all five stages of rising within four hours. The success of sourdough bread depends not only on the quality of the flour but is also a question of warmth.

November 2009

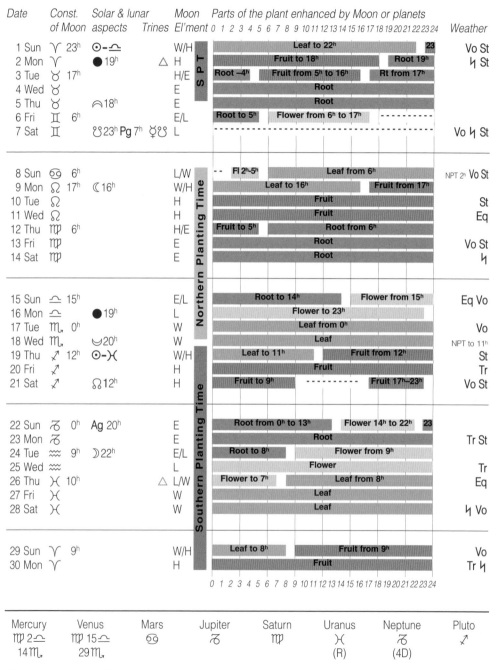

Date	Const. of Moon	Solar & lunar aspects	Trines	Moon El'ment	Parts of the plant enhanced by Moon or planets	Weather
1 Sun	♈ 23ʰ	☉–♎		W/H	Leaf to 22ʰ · 23	Vo St
2 Mon	♈	● 19ʰ	△	H	Fruit to 18ʰ · Root 19ʰ	♄ St
3 Tue	♉ 17ʰ			H/E	Root –4ʰ · Fruit from 5ʰ to 16ʰ · Rt from 17ʰ	
4 Wed	♉			E	Root	
5 Thu	♉	⌒18ʰ		E	Root	
6 Fri	♊ 6ʰ			E/L	Root to 5ʰ · Flower from 6ʰ to 17ʰ	
7 Sat	♊	☊23ʰ Pg 7ʰ ☿☊		L		Vo ♄ St
8 Sun	♋ 6ʰ			L/W	Fl 2ʰ–5ʰ · Leaf from 6ʰ	NPT 2ʰ Vo St
9 Mon	♌ 17ʰ	☾16ʰ		W/H	Leaf to 16ʰ · Fruit from 17ʰ	
10 Tue	♌			H	Fruit	St
11 Wed	♌			H	Fruit	Eq
12 Thu	♍ 6ʰ			H/E	Fruit to 5ʰ · Root from 6ʰ	
13 Fri	♍			E	Root	Vo St
14 Sat	♍			E	Root	♄
15 Sun	♎ 15ʰ			E/L	Root to 14ʰ · Flower from 15ʰ	Eq Vo
16 Mon	♎	● 19ʰ		L	Flower to 23ʰ	
17 Tue	♏ 0ʰ			W	Leaf from 0ʰ	Vo
18 Wed	♏	☽20ʰ		W	Leaf	NPT to 11ʰ
19 Thu	♐ 12ʰ	☉–⚹		W/H	Leaf to 11ʰ · Fruit from 12ʰ	St
20 Fri	♐			H	Fruit	Tr
21 Sat	♐	☊12ʰ		H	Fruit to 9ʰ · Fruit 17ʰ–23ʰ	Vo St
22 Sun	♑ 0ʰ	Ag 20ʰ		E	Root from 0ʰ to 13ʰ · Flower 14ʰ to 22ʰ · 23	
23 Mon	♑			E	Root	Tr St
24 Tue	♒ 9ʰ	☽22ʰ		E/L	Root to 8ʰ · Flower from 9ʰ	
25 Wed	♒			L	Flower	Tr
26 Thu	♓ 10ʰ		△	L/W	Flower to 7ʰ · Leaf from 8ʰ	Eq
27 Fri	♓			W	Leaf	
28 Sat	♓			W	Leaf	♄ Vo
29 Sun	♈ 9ʰ			W/H	Leaf to 8ʰ · Fruit from 9ʰ	Vo
30 Mon	♈			H	Fruit	Tr ♄

0 1 2 3 4 5 6 7 8 9 10 11 12 13 14 15 16 17 18 19 20 21 22 23 24

Northern Planting Time (SPT)
Southern Planting Time

Mercury	Venus	Mars	Jupiter	Saturn	Uranus	Neptune	Pluto
♍ 2♎	♍ 15♎	♋	♑	♍	♓	♑	♐
14♏	29♏				(R)	(4D)	

NB: All zodiac symbols refer to astronomical constellations, not astrological signs (see p.9)

Planetary aspects

*(**Bold** = visible to naked eye)*

1 $\mathbb{D} \, \text{°} \, ♀ \, 10^h$
2 $\mathbb{D} \, \text{°} \, ☿ \, 16^h$
3 $♀ \triangle ♆ \, 23^h$
4
5 $☉ \, \sigma \, ☿ \, 8^h$
6
7

8
9 $\mathbb{C} \, \sigma \, ♂ \, 4^h \quad \mathbb{C} \, \text{°} \, ♃ \, 18^h$
10 $\mathbb{C} \, \text{°} \, ♆ \, 3^h$
11 $☿ \triangle \, ♁ \, 12^h$
12 $\mathbb{C} \, \text{°} \, ♁ \, 5^h \quad \mathbb{C} \, \sigma \, ♄ \, 20^h$
13
14

15 $☉ \triangle \, ♁ \, 2^h \quad \mathbb{C} \, \sigma \, ♀ \, 16^h$
16
17 $\mathbb{D} \, \sigma \, ☿ \, 9^h$
18
19
20
21

22
23 $\mathbb{D} \, \text{°} \, ♂ \, 11^h \quad \mathbb{D} \, \sigma \, ♃ \, 20^h$
24 $\mathbb{D} \, \sigma \, ♆ \, 4^h$
25
26 $♀ \triangle \, ♁ \, 3^h \quad \mathbb{D} \, \sigma \, ♁ \, 13^h \quad ☿ \triangle \, ♂ \, 15^h$
27 $\mathbb{D} \, \text{°} \, ♄ \, 8^h$
28

29
30

Planet (naked eye) visibility
Evening: Mercury, Jupiter
All night:
Morning: Venus, Saturn, Mars

November 2009

Nov 8 2^h to Nov 18 11^h.
Southern Planting Time
Oct 22 to Nov 5 11^h and Nov 19 0^h to Dec 2.

Flower days until Nov 16 are particularly suitable for **planting flower bulbs** and will produced good growth with strong colourful flowers.

If not already done in October all organic materials should be collected and made into **compost.** Applying the biodynamic compost preparations brings about rapid colonization with fungal growth and ensures a good transformation. Barrel preparation also helps to create good humus.

Orchard and **forest trees** can now be planted supported by horn manure spraying and an application of barrel preparation.

Christmas trees can be cut on Nov 6, 12, 19 (after 12^h), 20, 21, 29 (after 10^h) and on Nov 30.

December 2009

All times in GMT

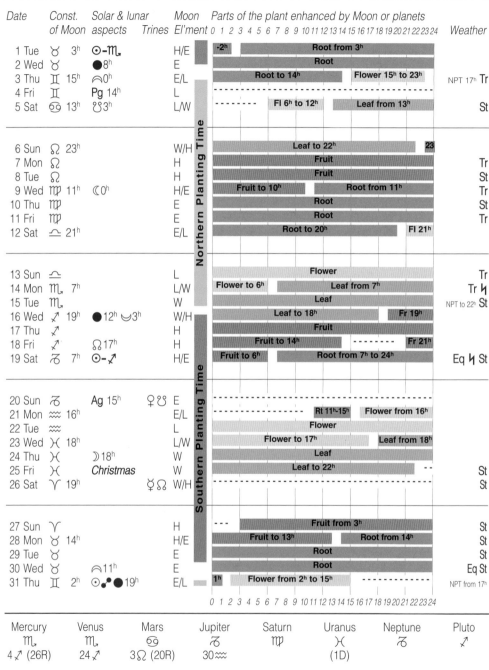

Date	Const. of Moon	Solar & lunar aspects	Moon Trines	El'ment	Parts of the plant enhanced by Moon or planets (0–24)	Weather
1 Tue	♉ 3h	☉-♏		H/E	-2h Root from 3h	
2 Wed	♉	●8h		E	Root	
3 Thu	♊ 15h	♎0h		E/L	Root to 14h · Flower 15h to 23h	NPT 17h Tr
4 Fri	♊	Pg 14h		L		
5 Sat	♋ 13h	☊3h		L/W	Fl 6h to 12h · Leaf from 13h	St
6 Sun	♌ 23h			W/H	Leaf to 22h · 23	
7 Mon	♌			H	Fruit	Tr
8 Tue	♌			H	Fruit	St
9 Wed	♍ 11h	☽0h		H/E	Fruit to 10h · Root from 11h	Tr
10 Thu	♍			E	Root	St
11 Fri	♍			E	Root	Tr
12 Sat	♎ 21h			E/L	Root to 20h · Fl 21h	
13 Sun	♎			L	Flower	Tr
14 Mon	♏ 7h			L/W	Flower to 6h · Leaf from 7h	Tr ♄
15 Tue	♏			W	Leaf	NPT to 22h St
16 Wed	♐ 19h	●12h ☋3h		W/H	Leaf to 18h · Fr 19h	
17 Thu	♐			H	Fruit	
18 Fri	♐	☊17h		H	Fruit to 14h · Fr 21h	
19 Sat	♑ 7h	☉-♐		H/E	Fruit to 6h · Root from 7h to 24h	Eq ♄ St
20 Sun	♑	Ag 15h	♀☊	E	Rt 11h-15h · Flower from 16h	
21 Mon	♒ 16h			E/L		
22 Tue	♒			L	Flower	
23 Wed	♓ 18h			L/W	Flower to 17h · Leaf from 18h	
24 Thu	♓	☽18h		W	Leaf	
25 Fri	♓	Christmas		W	Leaf to 22h	St
26 Sat	♈ 19h		☿☊	W/H		St
27 Sun	♈			H	Fruit from 3h	St
28 Mon	♉ 14h			H/E	Fruit to 13h · Root from 14h	St
29 Tue	♉			E	Root	St
30 Wed	♉	♎11h		E	Root	Eq St
31 Thu	♊ 2h	☉•●19h		E/L	1h Flower from 2h to 15h	NPT from 17h

Northern Planting Time · Southern Planting Time

Mercury	Venus	Mars	Jupiter	Saturn	Uranus	Neptune	Pluto
♏	♏	♋	♑	♍	♓	♑	♐
4♐ (26R)	24♐	3♌ (20R)	30♒		(1D)		

NB: All zodiac symbols refer to astronomical constellations, not astrological signs (see p.9)

Planetary aspects

(Bold = visible to naked eye)

1 $\mathbb{D} \, \text{☌}\text{♂}\text{♀}$ 14ʰ
2
3 $\mathbb{C} \, \text{☌}\text{♂}\text{☿}$ 10ʰ
4
5

6
7 $\mathbb{C}\sigma\text{♂}$ 0ʰ $\quad \mathbb{C}\,\text{☌}\text{♂}\,\text{♃}$ 5ʰ $\quad \mathbb{C}\,\text{☌}\text{♂}\,\text{♆}$ 9ʰ
8
9 $\mathbb{C}\,\text{☌}\text{♂}\,\hat{\text{☊}}$ 10ʰ
10 **$\mathbb{C}\sigma\hbar$ 5ʰ**
11 $\odot\triangle\text{♂}$ 1ʰ
12

13
14
15 $\mathbb{C}\,\sigma\,\text{♀}$ 22ʰ
16
17 $\text{♀}\triangle\text{♂}$ 13ʰ
18 $\mathbb{D}\,\sigma\,\text{☿}$ 8ʰ
19

20
21 $\mathbb{D}\,\text{☌}\text{♂}\,\text{♂}$ 3ʰ $\quad \text{♃}\,\sigma\,\text{♆}$ 10ʰ $\quad \mathbb{D}\,\sigma\,\text{♆}$ 12ʰ $\quad$ **$\mathbb{D}\,\sigma\,\text{♃}$ 12ʰ**
22
23 $\mathbb{D}\,\sigma\,\hat{\text{☊}}$ 23ʰ
24 $\mathbb{D}\,\text{☌}\text{♂}\,\hbar$ 21ʰ
25
26

27
28
29
30
31 $\mathbb{D}\,\text{☌}\text{♂}\,\text{♀}$ 15ʰ

Planet (naked eye) visibility

Evening: Mercury, Jupiter
All night:
Morning: Venus, Saturn, Mars

December 2009

Mercury, Venus and Mars are all in Water signs at the beginning of the month. However, on Dec 3 Mars moves into Leo and on Dec 4 Mercury enters Sagittarius and is joined at Christmas by Venus. On Dec 30 Jupiter moves into Aquarius where it will remain throughout 2010. At the end of the year only Saturn in Virgo and Neptune in Capricorn are in Earth signs. It is questionable whether these two on their own can cause really cold weather.

Northern Planting Time
Dec 3 17ʰ and Dec 15 22ʰ
and Dec 31 17ʰ to Jan 12.
Southern Planting Time
Nov 19 to Dec 2 21ʰ
and Dec 16 9ʰ to Dec 30 3ʰ.

The planting time is a good period for **pruning trees** and **hedges.** Fruit and Flower days should be chosen for pruning fruiting plants.

Christmas trees for personal use can be cut on Dec 22 or 23.

Burning of feathers and skins of warm-blooded pests should be carried out on Dec 1 between 3ʰ and 14ʰ. The full burning process must be completed by 14ʰ!

**We would like to wish all our readers
a blessed Christmas
and good health and strength of purpose
for the New Year**

Animal and insect pests

The first question to ask is: Why does a particular animal appear as a pest? The first step is to familiarize oneself with the living conditions and habits of the creature in question and to rectify any management errors that have been made. If despite this an animal becomes a pest, it can be driven back to within its natural limits using its own burnt remains — its ashes. For detailed recommendations see the monthly notes.

There is no need to reach for biological and chemical means of control; instead regulate the animal or insect from within its own species. For mice, birds and the like, a few skins or the feathers should suffice while for insects, slugs etc. the following method should be followed:

Take 50 or 60 specimens of the particular pest and burn them in a wood fire during the appropriate planetary aspect. The resulting mixture of wood and pest ash should then be "dynamized" by grinding it in a mortar for an hour.

One gram of this dynamized ash should then be placed in a small bottle with 9 cc (grams) of water and shaken vigorously for three minutes. This is the first decimal potency (D1 or X1). A further 90 cc of water are then added and it is again shaken for three minutes. This is the second decimal potency, D2 or X2. Repeating this procedure until D8 (X8) would produce 100,000 litres! It is therefore advisable to proceed until D4 and then start again using smaller quantities (1 cc of D4 in 9 cc water).

This D8 potency has been found to exhibit an inhibiting effect on the reproductive capacity of the pest when it is applied as a fine mist for three evenings in succession. Good results have been reported for several species. Where pests occur in large numbers good results are obtained by burning them on the site where they have been found. Flea beetle and apple blossom weevil can be caught with fly papers for example and burnt on site.

Concerning the ashing of wild animal skins

We always have many questions about this topic from our readers. In the woodland rich district where we live, the incursion of wild animals is a continuing threat to our trial beds. If fences are not dug at least half a metre into the ground, wild boar can easily lift them up while the roe and red deer simply leap over the top.

Burning skin in the wood oven   *Dynamizing the ash*   *Burning in the field*

A huntsman friend of ours gave us a skin. We burnt it in the wood oven at the time recommended in the calendar. The ash from the skin was then ground up together with the wood ash and dynamized for one hour using a pestle and mortar. A cement mixer can also be used. The skin ash was then potentized to D8 with the wood ash.

On one field the ash was sprinkled along the perimeter by hand. For a larger area Matthias put the ash into a small sowing machine. Using a piece of rolled up paper he set the machine so that only a minute amount of ash was released. In this way all the fields could be sprayed in a continuous line without any break. We then compared the effect of dynamized ash with the D8 potency.

In both cases the animals remained away from our cultivated fields. The effect of the deer ash could be clearly observed on an unfenced clover field. The animals had grazed the clover in the surrounding fields but not within two metres of the trial area. They had not crossed the line marked by the ash. Indeed the ash radiated its effect two metres beyond it.

To dynamize the ash we take 100 g (4 oz) of skin ash and 900 g (36 oz) of wood ash and grind it for one hour or beginning with 10 g (1/3 oz) we potentize it to D8 with either wood ash or water. For the dry material we use the sowing machine. When mixed with water we use a knapsack sprayer and on large areas a tractor-mounted sprayer. We had far better results using a machine than when we simply sprinkled it by hand.

Sowing times for trees and shrubs

Jan 24	Birch and Lime	Aug 1	Mirabelle Plum, Sweet Chestnut, Damson
Feb 5	Juniper and Blackcurrant		
March 8	Ash, Cedar, Pine, Juniper, Hazel	Aug 14	Ash, Apple, Hazel, Rowan
		Aug 20	Apricot, Peach
March 19	Apricot, Peach, Horse Chestnut, Plum, Sloe	Aug 21	Larch, Polar, Sallow, Snowberry, Lilac
April 5	Oak, Yew, Apple, Maple, Chestnut, Wild Rose	Aug 26	Mulberry, Walnut, Spindle Tree, Alder
June 23	Ash, Cedar, Fir, Hawthorn	Sep 11	Birch, Pear, Lime, Maple
July 4	Apricot, Peach	Sep 15	Juniper, Blackcurrant, Magnolia, Lilac, Thuja, Plum
July 30	Sallow, Damson, Sweet Chestnut	Sep 17	Fir, Ash, Rowan
July 31	Quince, Alder, Elder	Sep 23	Larch, Poplar, Sallow, Snowberry, Lilac

The above given dates refer to sowing times when the seeds of trees and shrubs are put in the earth. They are not times for transplanting already existing plants. The dates given are based on planetary aspects, which create particularly favourable growing conditions for the species in question. We then found that seeds from wild fruit species often produced fruit very similar to those of cultivated varieties but sweeter and with a stronger aroma.

Felling times for special timber

Oct 20	Larch, Lime, Elm, Horse Chestnut, Maple	Nov 15	Ash, Hazel, Spruce
		Nov 26	Oak, Yew, Chestnut, Walnut
Oct 29	Birch, Pear, Cedar, Copper Beech, Maple	Dec 11	Oak, Yew, Ash, Hazel, Spruce

Those trees which are not listed should be felled during November and December on flower days during the descending moon period (transplanting time).

The care of bees

A colony of bees lives in its hive closed off from the outside world. For extra protection against harmful influences, the inside of the hive is sealed with propolis. The link with the wider surroundings is made by the bees which fly in and out of the hive.

To make good use of cosmic rhythms, the beekeeper needs to create the right conditions in much the same way as the gardener or farmer does with the plants. The gardener works the soil and in so doing allows cosmic forces to penetrate it via the air. These forces can then be taken up and used by the plants until the soil is next moved.

When the beekeeper opens up the hive, the sealing layer of propolis is broken. This creates a disturbance as a result of which cosmic forces can enter and influence the life of the hive until the next intervention by the beekeeper. By this means the beekeeper can directly mediate cosmic forces to his bees.

It is not a matter of indifference which forces of the universe are brought into play when the the hive is opened. The beekeeper can consciously intervene by choosing days for working with the hive that will help the colony to develop and build up its food reserves. The bees will then reward the beekeeper by providing a portion of their harvest in the form of honey.

Earth-Root days can be selected for opening the hive if the bees need to do more building. Light-Flower days encourage brood activity and colony development. Warmth-Fruit days stimulate the collection of nectar. Water days are unsuitable for working in the hive or for the removal and processing of honey.

Since the late 1970s the varroa mite has affected virtually every bce colony in Europe. Following a number of comparitive trials we recommend burning and making an ash of the varroa mite in the usual way. After dynamizing it for one hour, the ash should be put in a salt-cellar and sprinkled lightly between the combs. The ash should be made and sprinkled when the Sun and Moon are in the Bull.

*A bee visiting
a foxglove
and clover*

Monthly notes for beekeepers

January: Mercury, Jupiter and Neptune are in Capricorn. This will ensure continuing winter dormancy. From Jan 22 cleansing flights will resume. Be sure to remove varroa packs in good time and keep the hive entrances clear.

February: Planetary influences will ensure that February remains a genuine winter month this year. Winter hive protection should not be removed yet.

March: Mercury and Mars in Aquarius from the second week onwards will encourage early flowering plants and provide for significant amounts of pollen. New brood activity will now be able to start.

April: Rain will dominate the first week. Mercury enters Aries on April 8 and should then bring warmth to the soil and warmer nights to support a good start to the bees' year. Bee plants should be sown on free areas.

May: Water and Earth constellations dominate in May and this could lead to a good harvest of honey due. Mercury and Mars will encourage honey production when they enter Aries on May 29 and 30 respectively. Continue planting bee plants wherever possible.

June: June should be a really good bee month with Venus and Mars in Aries and Jupiter and Neptune in Aquarius. It's a good time to expand colonies and breed queens. This should be done now since colder nights are likely in July. Use Flower days whenever possible.

July: Five planets are in Earth constellations during July and will bring cooler nights. The bees will now be able to rely on the flowers sown during April and May. Colony expansion should be completed.

August: Until Aug 19 the month will be very bee friendly. After that a lot of mist and fog is likely. Feeding for the winter should be concluded in good time so that the colonies can spend a fine September quietly preparing for the winter period.

September: From Sep 10 the month will show its sunny side. This will become even stronger from the middle of the month due to several planetary oppositions. Saturn will unfortunately move from Leo and spend the next four years in Virgo where he will only be able to support the bees' comb building drive.

October: During the first week bees will still be able to go on pollen forays. Later on October will present them with its cold shoulder.

November: The cold October weather will continue and the hives should now be made ready for winter.

December: On Dec 3 Mars enters Leo and Mercury, Sagittarius. We shall see whether they can prevail over the other Water-Earth constellations. It is important during changeable weather conditions that hive entrances are kept clear.

Keeping bees today

As one drives through the landscape in spring time one could be forgiven for thinking that the previously sad and flowerless landscape has been transformed into a bee paradise for, as far as the eye can see, there are vast fields of flowering rape giving off a rich scent. Looking more closely however we soon discover that hardly a bee is to be found. Only on certain days when sufficient rain has fallen and when the nights are warm enough can the rape fields provide a good flow of nectar for the bees. Interestingly the traditional rape growing areas of Europe were always close to the sea where a higher moisture level and more warmth could stimulate a greater flow of nectar.

In those areas bees, which were essential for pollination, produced unbelievably large amounts of honey which when properly processed was in great demand.

How should then honey be processed? With rape as with all flower honeys it is important to wait until most of the honey has been capped. To test this, a frame is

Landscape with rapeseed fields.   *Maize*

A frame of capped honey   *Extracting honey*

taken out and held horizontally and somewhat downwards to see if honey drips out. If it doesn't do so the honey can be harvested. It used to be believed that rape and dandelion honey should be harvested before it is capped because otherwise it would crystallize in the cells and be impossible to extract. This idea was probably spread by beekeepers who did not want to remove the caps before extraction. If good quality honey is wanted, however, it is vital that cell capping is as nearly complete as possible so that the qualities in the bee cells can take real effect in the honey.

After it is extracted, the honey should be allowed to settle for one or two days so that any air bubbles or pieces of wax can rise to the surface and be removed. Then comes the most important part of honey processing: stirring. A three-cornered piece of lime, beech or maple wood with an edge measuring 4 cm or 5 cm (2 in) can be used, or else a wooden stirring spiral.

While it is possible to use a machine for stirring, it is unfortunately often the case that when this machine is used the honey no longer crystallizes. This may be good for the user but it means that honey quality is reduced, for if honey can no longer crystallize it means that one of its key properties is lost.

After stirring with the wood or stirring spiral (one starts at the periphery and spirals in towards the centre) the honey will start forming fine crystals that dissolve like butter on the tongue and begin to reveal the diversity of flavours which good honey contains. It is then ready to store in a dark and cool room.

Let us move on from the rape fields and consider a plant which is more and more widely grown today, namely maize. As beekeepers we might be pleased about this additional source of pollen, even though we know that pollen from wind pollinated plants does not have the desired nutritional quality. But what do these vast areas of rape and maize give to us and the bees apart from honey and pollen? Huge quantities of pesticides are used, for without them the crops could not grow. Today these pesticides are of course harmless to bees, but what does that mean? When the foraging bee comes into contact or takes them up with the nectar or dew, it is not killed. What is the long term effect of these very small traces of pesticide, however? What effects do they have on the brood and on succeeding generations? Are the modern bee diseases connected? To this day none of these questions have been answered. But we do know that more and more people are keen to obtain food without pesticide residues. This has led to a rapid increase in the number of organic farms. Bee-keepers, too, need to find sources of nectar free of all chemicals so as to ensure the future health of their bees.

We continually receive questions about the possible effects of GM plants on bees. We cannot give a conclusive answer to this, however. Each person needs to develop their own understanding of the problem and form their own pictures. If we keep our eyes open and take note of less obvious media reports we will realize that there are very competent people out there with the courage to speak out in public about the damaging effects of genetic modification on plants and animals. They demonstrate that all such playing around with the genes of plants and animals should really cease. We can well imagine how such maize when fed to pigs causes infertility, and that if its pollen is used by bees their fertility is unlikely to improve and will probably decline. As they enjoy a juicy burger from an infertile pig, human beings will also absorb into their organisms the weakness of quality which led the pig to be infertile. If we want to have healthy bees we must offer them genuine and healthy plants.

Rudolf Steiner pointed out during his lectures on evolution how very close the connection is between humans and bees. If this is taken seriously the parallels between human beings and bees will not be hard to find. We will then soon acknowledge without need of further scientific proof that the environmental problems facing bees are the same as those facing us.

Albert Einstein stated in this connection: "If bees disappear from the earth, human beings will only be able to survive for a further four years. No bees, no pollination, no plants, no animals, no human beings."

The planetary influences of Neptune ♆ and Pluto ♇

Some specific observations of events are shared here which might be of interest to the reader. On the calendar pages **Eq** means a tendency towards earthquakes. For several decades we have been able to observe that earthquakes tend to occur when other planets are in a certain aspect to Neptune. Recently, while considering the earthquake pattern over the last forty-five years we calculated that during that period almost one million people lost their lives. On each occasion one planet or another stood at $90° - 45° - 135°$ or $72° - 144° - 36°$ to Neptune.

Vo in the calendar pages indicates a tendency for volcanic activity. When planets come into the above mentioned aspects with Pluto, volcanoes start to rumble. There are currently 500 to 600 active volcanoes on the earth which are generally quiet but can then erupt suddenly. Thus in February 2005 such large quantities of ash were spewed into the atmosphere in eastern Russia that scheduled air flights had to be cancelled. In March 2005 the same thing happened in North America and Mexico where ash ascended 11 000 metres (36 000 feet).

Once during a course we were running in Sicily we spent a night with a friend in Catania. In the middle of the night I woke up to find my room glowing red. I was convinced that something was burning and, startled, went and awoke my host. She responded by quietly telling me to look in my own Calendar. The Italian version clearly stated that volcanic activity was forecast for this date! Mount Etna continued to give the landscape a red glow even though no lava flowed.

During another visit we wanted to climb up Mount Etna and view the crater from above. After driving two thirds of the way up however we had to turn back because Etna had become active and a lava stream was flowing down through an orchard and across the road we would have had to pass.

While collating records spanning ten years, we found that of the 126 recorded volcanic eruptions, 106 occurred during the above mentioned Pluto aspects, most of the remainder were linked to earthquake activity, leaving only five which could not be explained. Volcanoes that are in continual activity were not included.

Unfortunately the International Astronomical Union decided last year that Pluto is not a planet and as a result it is no longer included in some more recent astronomical ephemerides.

Working with the biodynamic preparations

In lectures given at the beginning of the twentieth century, Rudolf Steiner spoke of how the earth is dying and will become a corpse and that human beings will need to bring it to life again. Among the audience were a number of farmers who subsequently asked: "What can we as farmers do to help?" Rudolf Steiner promised to give a course on agriculture, which he then did in June 1924. We have often referred to this in our calendar and other publications.

The biodynamic movement was born as a result of the recommendations given in those lectures. The compost preparations were created expressly to re-enliven the earth. To make them, several require carefully chosen animal organ materials which are taken and filled with certain medicinal herbs and placed in the earth during winter. In spring they are taken out and used to inoculate manure and plant compost which is then used to enliven the soil and so enable the plants growing in it to make better use of cosmic influences.

In England there was the widespread practice of feeding meat to ruminants. The animals got ill, had to be slaughtered and were then incinerated. The condition was called BSE and there was a great fear that the disease might be transferred to humans. Soon afterwards BSE appeared on the continent of Europe too and the EU Commission decided that animal organ materials were a risk and could no longer be used. Butchers were instructed that all the internal organs of cattle should be immediately destroyed. This of course affected biodynamic farmers particularly strongly. After all what is biodynamic agriculture without the compost preparations? We had worked with them for many decades and carried out numerous tests on them.

We then set out to discover whether the organs of sheep, goats, or deer could achieve the same effect as those from cows. One would of course have needed far more animals since their organs would be much smaller. During earlier research we had discovered that the yarrow preparation mediates Venus forces, chamomile Mercury forces, oak bark Mars forces, and dandelion Jupiter forces. Through our long term involvement with trees we were also aware of their respective links to planets. We therefore decided to find out whether compost preparations made using wood from these trees would be as effective as those made in the normal way. This meant of course that in the following year we had to make and test an endless number of compost piles! Afterwards we set up trials with many different plants including a neutral follow up planting so that the reproductive strength of the affected plants could be investigated. Further trials were needed beyond this, however, in order that through analytical comparisons we could find out whether we were going in the right direction. In a new publication about the biodynamic preparations due out soon, we will present a much fuller description so that practitioners can make their own experiments.

Four diagrams from the manuring trials of 2006

In 2004 we made the biodynamic compost preparations in two different ways. We made one set in the classical way using animal organ materials as covers. A second set we made by replacing the animal organs with sheaths from trees and using these for the compost. After doing this for two years we could begin to compare results.

These diagrams show the yields of phacelia, corn salad, spinach and spinach seeds. It was amazing to see that in the case of both horse manure and cow manure the highest yields arose when they were treated with the so-called vegetarian preparations. This was very comforting to us in the face of current restrictions in the EU on the use of animal by-products. The four diagrams show these positive yield effects and could be a useful means for enabling farmers and gardeners affected by these restrictions to continue working biodynamically.

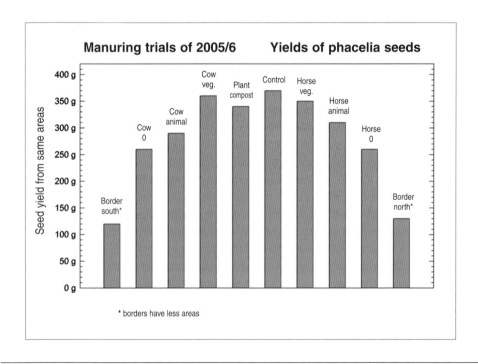

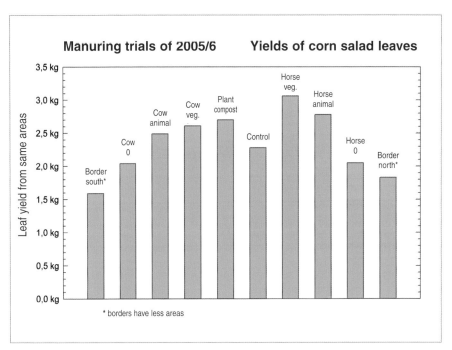

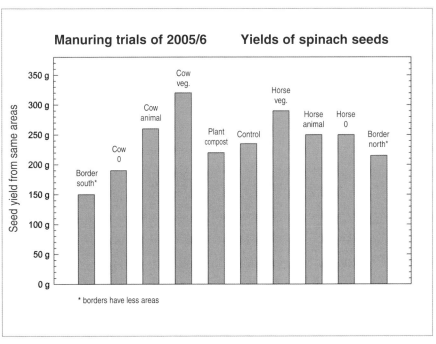

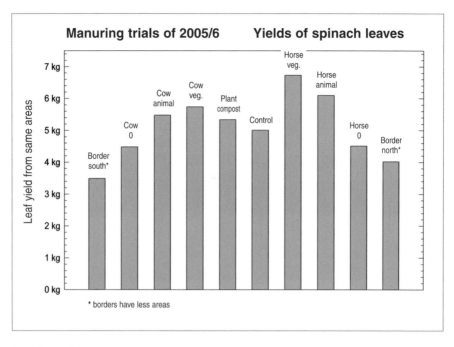

Manuring trials of 2005/6   Yields of spinach leaves

Leaf yield from same areas

- 7 kg
- 6 kg
- 5 kg
- 4 kg
- 3 kg
- 2 kg
- 1 kg
- 0 kg

Border south*
Cow 0
Cow animal
Cow veg.
Plant compost
Control
Horse veg.
Horse animal
Horse 0
Border north*

* borders have less areas

Trials of 2008

On the Heckacker field we ran trials during planetary oppositions. Taking radishes and spinach as test plants. Before sowing soil samples were taken for testing. For each of the twenty sowings manure preparation has to be stirred, which is then applied in three sprayings.

Recipe for rye bread

Since all the grains we grow in our experiments were tested for quality, we have used our own raising agents (without additives) for bread baking to produce a good loaf. Besides sour milk, buttermilk, whey and syrup we have also tried to bake with honey and have developed a recipe that has proved its worth over many years. One heaped teaspoonful of flower honey is stirred well in a glass of warm water (50°C, 120°F) and then mixed with 250 g ($^1/_4$ lb) of finely-ground rye meal.

This small amount of dough is made in the evening and kept warm overnight. It should be at a temperature of about 26–30°C (79–86°F) by the stove or next to a hot plate, which is set very low. Next morning add the same amount of rye meal and warm water or whey. In the evening add the rest of the flour (approximately 60% of the total) to the prepared dough with sufficient warm water. At this stage you can add a little linseed, caraway, fennel or something similar and leave it to rise overnight. Next morning add salt and finish the dough. When it begins to rise again the loaves are formed. Let them rise well, put them in a preheated oven, and bake them for a good hour. Rye is easy to digest when it has gone through these five stages. You can keep about 500 g ($^1/_2$ lb) of the finished dough and leave it in an earthenware pot. After it has risen again a little, it should be sprinkled with salt, covered with grease-proof paper and stored in a cool place (not the fridge).

When you want to do some more baking, take the pot from the cool place in the morning and add a teaspoon of honey, which has been stirred in a glass of warm water. Then keep the pot warm. In the evening you can start on the main dough and proceed as described above. You can also begin from stage one again, but then it will take longer with this kind of sour dough. Rye should rise five times. Wheat, barley and oats need to rise only three steps. Success depends on the warmth of the baking area.

Readers' questions concerning potato and tomato culture

Forty years ago plant breeders focused their energies on developing better potatoes. They were, however, concerned less with strengthening the whole plant than making better use of nitrogen fertilizers. Potato health was sacrificed to fertilizers. I still remember as a child how potato land was treated with manure and compost in the autumn and how my father insisted on planting and hoeing only during fine weather.

In 1963 we began potato trials using cosmic rhythms and biodynamic preparations. On the areas designated for potatoes in our crop rotation we gave an autumn application of 2 tons of well rotted manure compost. The compost is best worked into the surface of the soil in September together with a spraying of barrel preparation. The ground should be ploughed and ready for the winter by the end of October. A further application of barrel preparation can accompany the plough.

As soon as the soil warms up in spring it is cultivated on a Root day and sprayed with horn manure and barrel preparation. Then the potatoes are planted. In colder areas it is best to wait until the Sun has moved into Taurus (May 13). The potatoes will then germinate quickly and grow strongly. After about four weeks the potatoes should be sprayed twice in the evenings with stinging nettle tea on Leaf days.

Horn silica should be sprayed early in the morning on Root days three times during the following weeks. All the necessary cultivations including ridging up should always be carried out on Root days. If in the autumn the potato tops have not died down sufficiently, an application of horn silica can be given on Root days in the afternoon. The harvest should also be carried out on Root days. With this treatment we have never had any problems with phytophtera, potato blight, and obtain good yields.

Some growers have problems with the Colorado beetle. The beetles should be collected up and stored in a glass jar with a lid until next spring. They should then be burnt when both Sun and Moon are in Taurus, potentized to D8 and sprayed three times when the Moon is in Taurus. It can also be applied in the autumn on ground where the next season's potatoes will be grown when it is ploughed.

Tomatoes belong to the group of plants known as heavy feeders. It is, however, not an animal that needs feeding large quantities of manure. In an answer to a question during the Agriculture Course, Rudolf Steiner said that the tomato would prefer above all to grow in its own compost.

We have undertaken many different trials with tomatoes. During our research

Planting potatoes

Flowering potatoes

work we have however been unable to confirm this particular indication. We have gathered the tomato halms in the autumn and composted them using the biodynamic compost preparations and then used this compost for the tomato bed the following spring. We then compared the results with manure compost and general plant compost.

The tomatoes which had received the equivalent of 1 ton of manure compost in the autumn produced the best and healthiest plants and with very good yields. Those which had only received plant compost appeared under-nourished even though hoof and horn meal had been added to the compost. The plants grown in their own compost suffered fungal attacks already in early growth stages.

With every sowing whether in the greenhouse, on the window-sill or when planting outdoors, an application of horn manure is given.

In the crop rotation tomatoes can follow the cabbage family. As soon as the plants have their third pair of leaves they are given an evening spray of nettle tea. This is repeated three times during the course of one month. On the following fruit days horn silica is stirred early in the morning and sprayed on the plants, also three times within a month.

Side shoots should be removed on fruit days towards evening. All such work on the plants should be undertaken on fruit days. The fruit is best picked on fruit and flower days avoiding the unfavourable blanked out days.

To collect seed, fruit should selected by the end of September and not the last ones growing at the end of November. If, when the time comes for clearing away the plants, some good green fruits are still hanging, the whole plants can be pulled out

Planting potatoes

Flowering potatoes

of the earth with their roots and hung upside down in a sunny place. The fruits will then still ripen sufficiently to serve for supper.

With the triple applications of nettle tea and horn silica, the leaf surfaces will be strengthened to such an extent that no spores of blight will be able to gain a foothold. If there are late frosts after the tomatoes have been planted outside, 10 drops of valerian can be stirred for ten minutes in ten litres ($2^1/_2$ gal) of water and sprayed on them. The leaves usually hang wilted for a while afterwards. A spraying of stinging nettle tea should then be given to both plants and soil followed by a thorough watering with clear fresh water. After a few hours the plants will be standing upright again and grow on normally.

Horn silica should be sprayed on top fruit and berries during summer in order to support bud development for the following year on June 22, 26 and 27 June and on July 2, 6 and 7.

Concerning microwave radiation (from readers' questions)

From the work of Prof Ewald Schnug und Frau Liedemann we discovered a lot about the damage caused by microwave radiation. The installation of a phone mast in our neighbourhood close to our trial fields brought the issue into our immediate awareness. Many experts have been working on this problem but have until now found no solution. I have experienced that hazelnut trees and shrubs can help to hold back this radiation as can rye straw and peat. It is questionable today whether any part of the landscape is free of such radiation.

The rose family (Rosaceae)

Most of our fruit belong to the rose family. Like all plants they have a close relationship with particular planets. In order to harvest a good crop of fruit it is important to note the following: Fruit bud development takes place the previous summer. If the planetary ruler of the particular fruit variety is in a Fruit constellation during this period this will be the best time for carrying out summer pruning or applying the horn silica preparation.

Roses belong to Mars. Those years when Mars is in flower constellations during the relevant weeks of summer should be made particular use of by rose enthusiasts. An autumn application of stinging nettle compost followed by a spring spraying of stinging nettle tea will further support prolific flowering in the following year.

A somewhat different approach to quality

Ever since 1952 when we began our research into cosmic rhythms and the bio-dynamic preparations, we have sought to produce food of the highest quality while at the same time doing whatever is best for the earth.

Today apart from conventional farming practices there are organic and ecological ones. The latter do not apply heavy chemicals, herbicides etc., many follow the recommendations given in this calendar and even plant teas are sprayed to enhance quality. All that is missing to attain Demeter quality is the application of biodynamic preparations.

There are animals that know the value of Demeter quality very well, for example wild pigs. They know where Demeter potatoes or wheat are to be found in a given district. Once, I asked the local hunters if they could shoot more wild pigs in the area. They replied that they had shot over 150 animals in the neighbourhood of just three villages. They also mentioned how the sows always seemed to visit our fields first, which suggests that they have a nose for Demeter quality!

Above: A family of wild pigs

Right: Wheatfields destroyed by wild pigs

Above: Rye field destroyed by deer

Left: Stag

They can only be kept out by digging fences half a metre (18 in) into the ground, fixing barbwire directly above the soil and using an electric fence.

Roe deer and red deer also know where to find Demeter rye. An electric fence doesn't stop them; they simply jump over it. To stop roe deer the fence must be 2.5 m (8 feet) high and for red deer even higher. The animals test the height of the fence and before you know it they are over and enjoying the rye.

Our hens also like Demeter grain. If they are fed Demeter wheat they put on fat, if they are fed Demeter oats they lay more eggs and of course produce more egg shells for us. The resulting high quality of our hens is recognized by the fox too. One day he took four of our hens. We assumed that he would only eat one of them and so went to find out where he had hidden the others. We didn't find them but instead found the fox dead by the roadside. Perhaps he had consumed so much Demeter quality that he was no longer fast enough to avoid the traffic!

Sheep *Thistles* *Cows*

Between the rye and the wheat fields lay a strip of "Demeter thistles" which neither the wild pigs nor the deer had touched. They were later taken when a shepherd brought his sheep and goats through it. They ate the still flowering thistles right back to the main stem.

I remember from my childhood how we used to deal with thistles. One very dry year it hadn't rained for many months and in those days it wasn't possible to buy animal feed. The cows in the village no longer produced any milk. Each evening my father and I went with an oxcart to areas of rough ground and along the woodland edges. There we collected thistles and nettles and brought them home. They were put in a big pan with a lot of chaff and boiled. The mixture was left to stand overnight and in the morning was fed to the cows. As a result the cows gave a good quantity of milk with a high fat content, much more in fact than from normal fodder. Our farm was the only one in the village, which could still sell butter on the market. The neighbours were amazed at this and asked me what we fed our cows on. I told them, with thistles and nettles. But they didn't believe me and thought it was a joke!

Moon diagrams

The diagrams overleaf show for each month the daily position (evenings GMT) of the Moon against the stars and other planets. For viewing in the southern hemisphere, turn the diagrams upside down.

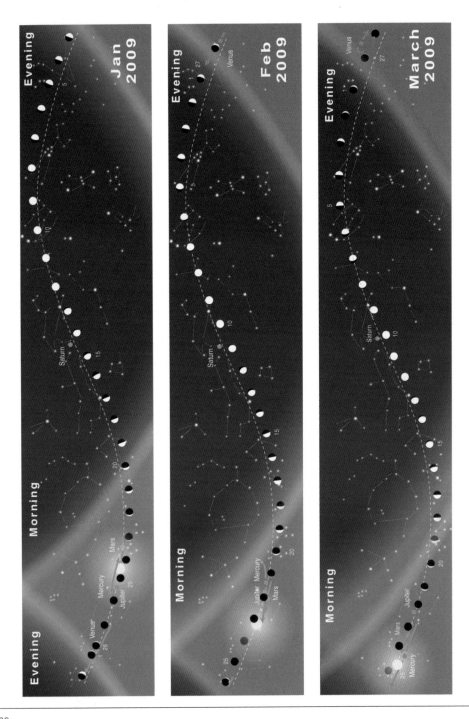

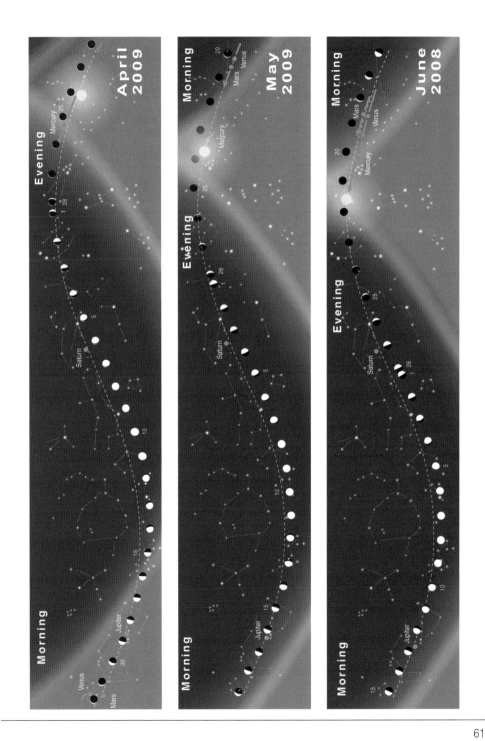

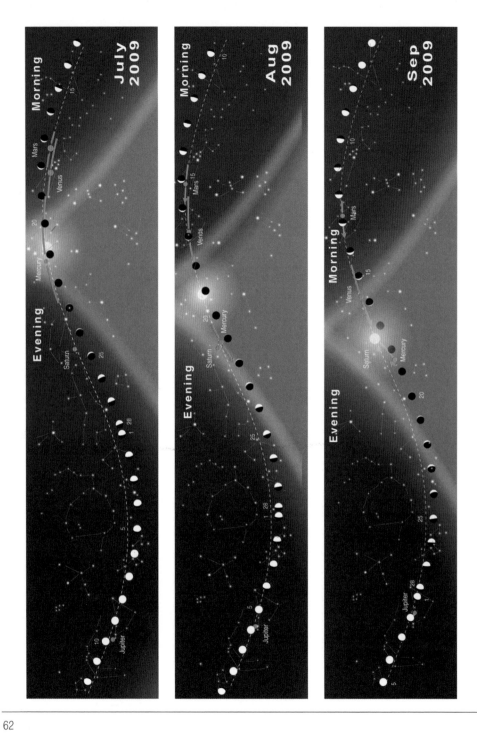

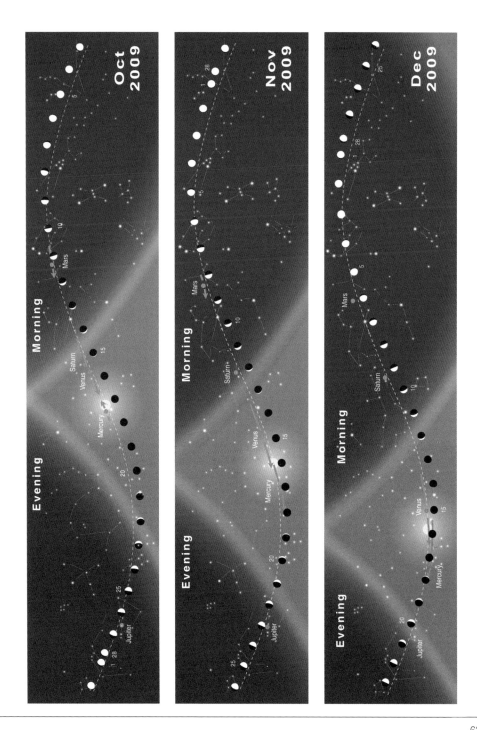

Further reading

Bockemühl, Joachim, *In Partnership with Nature*. Anthroposophic, USA.
—, *Extraordinary Plant Qualities for Biodynamics*, Floris.
Cloos, Walther, *The Living Earth*, Lanthorn.
Colquhoun, Margaret and Axel Ewald, *New Eyes for Plants*, Hawthorn.
Conford, Philip, *The Origins of the Organic Movement*, Floris.
Grotzke, Heinz, *Biodynamic Greenhouse Management,* Biodynamic Literature, USA.
Klett, Manfred, *Principles of Biodynamic Spray and Compost Preparations,* Floris.
Koepf, H.H. *The Biodynamic Farm*. Anthroposophic, USA.
—, *Research in Biodynamic Agriculture*, Biodynamic Farming & Gardening Ass. USA.
Kranich, Ernst Michael, *Planetary Influences upon Plants*, Biodynamic Literature, USA.
Moore, Hilmar, *Rudolf Steiner's Contribution to the History and Practice of Agricultural Education*, Biodynamic Literature, USA.
Philbrick, John and Helen, *Gardening for Health and Nutrition*, Garber, USA.
Remer, N. *Laws of Life in Agriculture*, Biodynamic Farming & Gardening Ass. USA.
—, *Organic Manure*, Biodynamic Farming & Gardening Association, USA.
Sattler, F. & E. von Wistinghausen, *Biodynamic Farming Practice*, Biodynamic Agricultural Ass.
Schilthuis, Willy, *Biodynamic Agriculture*, Floris.
Soper, John, *Biodynamic Gardening*, Biodynamic Agricultural Ass.
Steiner, Rudolf, *Agriculture (A Course of Eight Lectures)*, Biodynamic Literature, USA.
Storl, Wolf, *Culture and Horticulture*, Biodynamic Literature, USA.
Thun, Maria, *Gardening for Life*, Hawthorn.
—, *Results from the Biodynamic Sowing and Planting Calendar,* Floris.
von Keyserlink, Adelbert Count, *The Birth of a New Agriculture*, Temple Lodge.
—, *Developing Biodynamic Agriculture*, Temple Lodge.
Weiler, Michael, *Bees and Honey, from Flower to Jar,* Floris.

Biodynamic Associations

Australia: Biodynamic Agricultural Association, PO Box 54, Bellingen, NSW 2454. *Tel:* 02 6655 0566. *Fax:* 02 6655 0565. *Email:* bdoffice@biodynamics.net.au *Web:* www.biodynamics.net.au

Canada: Demeter Canada, 115 Des Myriques, Catevale Que. J0B 1W0. *Tel:* 819-843-8488. *Email:* laurier.chabot@sympatico.ca *Web:* www.demetercanada.com

Ireland: Biodynamic Agricultural Association, The Watergarden, Thomastown, Co. Kilkenny. Tel/*Fax:* 056-54214. *Email:* bdaai@indigo.ie *Web:* www.demeter.ie

New Zealand: Biodynamic Farming & Gardening Association, PO Box 39045, Wellington Mail Centre. *Tel:* 04-589 5366. *Fax:* 04-589 5365. *Email:* biodynamics@clear.net.nz *Web:* www.biodynamic.org.nz

South Africa: Biodynamic and Organic Agricultural Association, PO Box 115, 2056 Paulshof. *Tel:* 011-803 1688 *Fax:* 011-803 7191.

UK: Biodynamic Agricultural Association (BDAA), Painswick Inn, Gloucester Street, Stroud GL5 1QG. Tel/*Fax:*01453 759501. *Email:* office@biodynamic.org.uk *Web:* www.biodynamic.org.uk

USA: Biodynamic Farming and Gardening Association, 25844 Butler Road, Junction City, OR 97448. *Tel:* 888-516-7797 or 541-998-0105. *Fax:* 541-998-0106. *Email:* biodynamic@aol.com *Web:* www.biodynamics.com